Wildflowers of the
Southern Mountains ⁊⁊

*To Elaine
with every good wish*

Richard M. Smith

Wildflowers of the Southern Mountains

Richard M. Smith

Illustrated by the Author

The University of Tennessee Press
Knoxville

First Edition.

Frontispiece: Gray's Lily. All photos and line draw-
ings were prepared by the author.

The paper in this book meets the minimum require-
ments of the American National Standard for Per-
manence of Paper for Printed Library Materials.
∞
The binding materials have been chosen for strength
and durability.

Library of Congress Cataloging-in-Publication Data

Smith, Richard M., 1914–
Wildflowers of the southern mountains / Richard
M. Smith; illustrated by the author. — 1st ed.
    p. cm.
Includes bibliographical references (p.) and index.
ISBN 0-87049-992-0 (pbk. : acid-free paper)
1. Wildflowers—Appalachian Region, Southern—
Identification. 2. Wildflowers—Appalachian
Region, Southern—Pictorial works.
I. Title.
QK122.3.S55 1998
582.13'0975—dc21   97-21013

Contents ❧

In memory of my mother
who would have loved this book
and my father
who could have done it better

Introduction ✤

For more than 150 years after the Europeans succeeded in establishing colonies on the southeastern coast of North America, they remained almost totally ignorant of the natural riches that lay in the mountains only a few hundred miles to the west. In the early 1700s Mark Catesby, an English naturalist, made tentative excursions into this unknown country to gather material for his *Natural History of America*, and a handful of trading posts had been set up. But it was not until the spring of 1775 that the botanist William Bartram of Philadelphia ventured into the land of the Cherokees to make the first determined effort to learn about Native Americans' way of life, the fauna of the region, and especially its flora.

Bartram was given to extravagant prose, but entries in his journal like the one in which he described "this magnificent landscape, infinitely varied, and without bound" were later found by others to be virtually free of exaggeration.*

As strange, newly discovered plants and their seeds were shipped by collectors to their patrons in the Old World, European botanists began to swarm over the region. Many have been commemorated in the names of now familiar species, among them André Michaux, John Fraser, Frederick Pursh, and Thomas Nuttall. This enthusiasm for the lavish flora of the southern mountains continued for decades, and, in the mid-1800s America's pre-eminent botanist, Asa Gray, proclaimed them to

---

*The Travels of William Bartram (Naturalist's Edition), ed. Francis Harper (New Haven, Conn.: Yale Univ. Press, 1958), 212.

be among his favorite plant-hunting grounds. Even to-day their diversity of plant life, unsurpassed anywhere except in the tropics, is famous throughout the world.

One remarkable feature of the southern Appalachian flora is its similarity to that of eastern temperate Asia; among many genera common to both regions are *Trillium* and *Shortia*. This phenomenon was of great interest to Gray, who attributed it to the thousand-mile-wide land-bridge that once connected Alaska and Siberia.

The concentration of such an astonishing variety of trees, shrubs, herbs, and other plant forms in the relatively small area encompassed by the southern Appalachians and their neighboring ranges can be ascribed to several factors. For one, these mountains are ancient, as attested by their worn, rounded contours in contrast to the sharp, jagged peaks of much younger chains. For 200 million years they stood high enough to avoid submersion beneath the sea, and they escaped the destructive effects of the continental glaciers that scoured the northern half of North America. This created a sanctuary in which the evolutionary development of plants was able to proceed without interruption.

Other contributing factors include rich soil, abundant rainfall, and fast-flowing streams. Mild temperatures at the lower elevations are favorable to species from the neighboring piedmont, while the cooler climate on the highest summits provides refugia for boreal plants that had been forced southward ahead of the advancing ice sheets.

To all this must be added the impact of human occupation, for many alien plants introduced into these mountains have competed well enough to make it on their own. Whether brought in deliberately or inadvertently, one effect has been to further increase the species diversity and botanical interest of our mountain flora.

In contrast to their remoteness from the early colonial settlements, the southern mountains are now accessible to an exceptional degree. The Blue Ridge Parkway, in combination with the Skyline Drive in Shenandoah National Park, provides 574 continuous miles of pleasurable driving through the highlands of Virginia and North Carolina, and other roads reach the tops of some of the highest peaks east of the Mississippi. At its southern end the Parkway meets Great Smoky Mountains National Park, the most popular in the entire federal park system and the ecological crown jewel

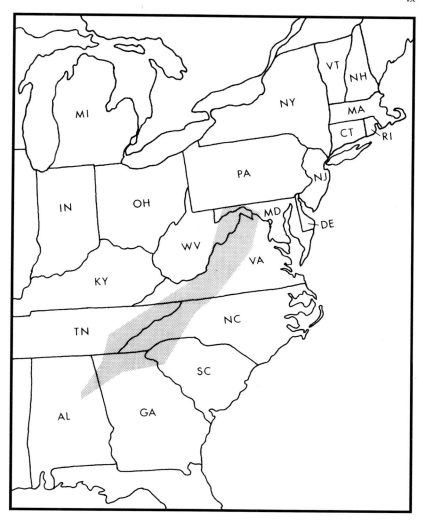

The Southern Mountains Region.

of the eastern mountains. The Appalachian Trail fol-
lows the crest line into Tennessee, reaching the heart
of the Smokies before turning south to its terminus in
Georgia. Many other trails and campgrounds are to be
found throughout the region within national forests,
in state recreational areas, and on private lands.

Although the term "southern mountains" has served
well enough for the purposes of the foregoing discus-
sion, broadly interpreted it implies a flora so multifari-
ous that it could not be treated except very superficially
in a book of this size. To avoid this all too common

predicament, it was decided to focus primary attention
on the physiographic segment that contains the richest
concentration of wildflower species and therefore best
exemplifies the whole. The territory selected is a complex that for the most part forms the eastern flank of
the southern Appalachians and includes the Blue Ridge
and Great Smoky Mountains, but also extends in places
into the Alleghenies and the piedmont. Indicated in a
general way by the shaded areas on the accompanying
map, this is the portion of the southern mountains for
which the representation of wildflower species in this
volume will be found to be essentially complete.

This work's usefulness will by no means be so narrowly circumscribed, however. Plants are not affected
by artificial boundaries, so a high percentage of coverage can also be expected for the contiguous prov-
inces such as the piedmont and ridge-and-valley; this
overlapping will diminish as one travels west and
south and—more gradually—toward the east and
north. In fact, a great many of the wildflowers occurring in the twenty-one-state region depicted by the
entire map will be found in these pages. (Interest-
ingly, some southern mountains have disjunct popu-
lations of species that are otherwise confined to New
England peaks or the Atlantic coast.)

The same practical considerations that limit the size
of a region the flora of which can be dealt with satisfac-
torily also impose other restrictions. For example, the
number of species described had to be held down to
around 1,200. Consequently, plants that do not conform to the usual perception of a wildflower have been
excluded: trees, shrubs, and woody vines; grasses, sedges,
and rushes; submersed aquatics; and several groups of
"weeds" that bear insignificant flowers. As with any rule
there are the inevitable exceptions, of course.

WHAT'S IN A NAME?

Seeking out and enjoying wildflowers is certainly one
of the most pleasurable and enlightening activities in
the field of natural history, and learning to know them
by name is the key to unlocking a vast storehouse of
knowledge concerning them. In fact, without knowing
the name of a plant we cannot even begin to search for
other information about it.

We might, for instance, want answers to questions

like these: Where in the world did this wildflower originate, how did it come to be here, and can I raise it at home? What peoples have used it for food, clothing, or medicine in ages past, and what does science know today about its current and potential value? What is its place in the environment, and how do its functions relate to the soil, the climate, the animate organisms with which it coexists, and the neighboring vegetation? Another important consideration: How is it getting along with humankind—is its place secure or is its survival in peril? In every case, our inquiry must start with the name by which the plant is known.

Some examples of answers to questions such as these will be found here, in a number of short paragraphs interspersed throughout the text (and differentiated from it to minimize intrusion into the identification process). Many people may be surprised at the extent to which plants have been used as medicine, for instance, and it might be worthwhile to digress a bit to consider why this is so. Actually, it requires little reflection to realize that, at the very beginning, the human race had almost nowhere to look for help in treating an illness or assuaging a hurt *except* to the world of plants—and this remains largely true even today. One group of Native Americans explains the relation between healing and plants by citing an ancient legend. All of the animals convened to discuss ways of defending themselves against the slaughter they were experiencing at the hands of human hunters. It was decided that each species of mammal would in turn visit upon humans a particular ailment; the deer said they would be responsible for inflicting rheumatism, for example. They were followed by the reptiles, birds, insects, and other creatures. When the plants, who were less aggrieved and therefore more disposed to be friendly, heard of this they agreed that every kind of tree, shrub, herb, or other plant should respond with a remedy whenever human beings should call upon them for help. This, then, was the origin of medicine.

We can only try to imagine the difficulties that must have attended early efforts to find effective cures among plants. Much reliance was placed upon supernatural forces for guidance, as in the Middle Ages when the so-called Doctrine of Signatures was given wide credence. This held that plants were marked by the Creator with clues to indicate their medical properties, so that the sorrels, which have heart-shaped leaflets, would cure cardiac diseases; likewise, the irregular convolutions on a walnut kernel, which resemble those on the brain, recommended it for treating mental disorders; and so on.

Despite such unscientific reasoning, it is of course true that we owe a great deal to "folk remedies," many of which have turned out to have legitimate medicinal properties—willow as a source of aspirin, and foxglove of digitalis, to name but two—but many beneficial plants are just as capable of causing death. Even some parts of otherwise edible plants, such as the leaf blades of rhubarb, are poisonous. It should, therefore, be an absolute rule never to attempt self-medication with any plant, or sample it as food, unless it has been positively identified by a qualified and responsible authority as being safe.

No matter how compelling the argument for learning the names of plants, however, the beginner should be aware that it is possible to spoil the fun by trying too hard. With nearly a quarter of a million kinds of flowering plants in the world, just being able to name those that grow in a small area can be challenging to all but the experts, and there are some groups, like the goldenrods, for example, about which even they find it impossible to agree. So do not expect too much of yourself at first. A walk of a few hours in the mountains may yield dozens of unfamiliar wildflowers, but if you manage to identify a mere handful and simply enjoy the rest it will have been a day well spent.

It cannot be stated too often or too emphatically that the only legitimate way to identify a plant is to learn its recognized *scientific* name. The common names that we have become accustomed to using are actually more confusing than helpful. In many cases multiple common names have been conferred upon a single species (often as a result of regional preferences), and in others the same name has been applied to a number of different and frequently unrelated species. Someone has counted more than a hundred kinds of plants called "Mayflower"—a striking example of a name having been rendered virtually meaningless by indiscriminate use. This is not to suggest that you will get any further with your complaint about the dandelions in your neighbor's lawn by calling them *Taraxacum officinale*, but save the vernacular for occasions when you are sure that others will know what you are talking about.

Scientific names are based upon a binomial system that traces its origins to the Europe of more than three centuries ago but was first consistently applied by the Swedish taxonomist Linnaeus in his *Species Plantarum* in 1753. Fortunately for us, this system did away with the earlier use of cumbersome polynomials, which really were strings of Latin words amounting to descriptive sentences and were much too lengthy to serve as names.

As the term indicates, a binomial consists of only two words: the name of the genus (potentially shared with a group of closely related plants) and a specific epithet, which is the unique and distinctive designation of one kind of plant within that group. These epithets are drawn from various sources; most often they are Latin- or Greek-derived adjectives describing the plant, but frequently they consist of the latinized name of a person or place. Examples of these can be seen in *Trillium grandiflorum*, *Chelone lyonii*, and *Rhexia virginica*.

Inevitably it became apparent that a set of rules would be needed to ensure that all countries would follow a uniform procedure in the selection of botanical names, and in 1867 the International Botanical Congress adopted what was known as the Paris Code. This formed the basis of the present International Code of Botanical Nomenclature, by means of which we are now assured that the accepted names are unique and universal, each species being called by its particular name, and only by that name, everywhere in the world regardless of local languages.

Scientific names are not, however, completely immune to change. The essential underlying principle of the International Code is that of priority, and when, for instance, it is discovered that a name being used is not in fact the oldest, the rules require that it be set aside in favor of the earlier one. Also, differences of opinion or interpretation occur more often than we might like. The nomenclature used herein reflects a number of recent changes, but a good many alternate scientific names which formerly appeared in other publications are supplied parenthetically in order to make things as easy as possible for the reader.

Plants are classified scientifically in a hierarchical arrangement of groups beginning with the species, which is the basic natural unit of classification. Its individual members are usually numerous and essentially identical in structure. They are capable of breeding among themselves but cannot interbreed with members of other groups and produce fertile offspring. Certain groups of species which have a closer resemblance to each other than to other species are combined into the next higher category, the genus. Similarly, genera are grouped into families, families into orders, and so on. Family names always end in the suffix "aceae," and although they are sometimes added parenthetically to the binomial they are not a necessary part of the plant name.

Sometimes a local population of plants will be found to exhibit minor morphological differences which, although not significant enough to justify separation of the taxon into independent species, may suggest that they should be accorded intraspecific rank. When this occurs, a taxonomist may choose to designate them as a subspecies (abbreviated "ssp."), a variety, or a form, and to identify it as, for example, *Viola pedata* var. *lineariloba*. Such a designation may be written in full if it is desired to differentiate between the two taxa; otherwise, the binomial *Viola pedata* is correct and adequate as a name for all Birdfoot Violets.

How to Use This Book

Think of the classification of the world's plants as an enormous, intricate tapestry woven by Linnaeus and his precursors and successors, in which each thread represents a single species and is juxtaposed with those which are similar in appearance and therefore presumed to be related.

Continuing the analogy, the task of identifying a particular plant species becomes that of reversing the process, methodically unraveling the fabric until we locate the one and only thread that matches our specimen. It is no wonder, then, that in every identification guide there needs to be some kind of helpful table or chart, appropriately called a "key."

The whole tapestry is much too large for a random hunt, so we must narrow our search to just that portion which coincides with our field of interest, and here it becomes important to select a guide book that 1) focuses on the right geographic region and yet 2) includes a high percentage of that region's wildflowers. This will increase the likelihood that whatever we find in the field probably will be found in the book.

Regardless of the sort of key that may be provided, our first inclination is to attack the problem head-on by riffling through the pages looking for a picture that resembles the specimen, then reading just enough text to confirm our diagnosis. There is much to be said for this method, especially when, as in this instance, nearly one-half of the species are illustrated by photographs which have been selected for their depiction of identifying characteristics. Beginners should realize, however, that in doing so they will be missing out on the most useful shortcut of all, which is the ability to go directly to the plant family and begin the search there. The patient use of a key will afford you the means of developing this faculty almost without realizing it, and, paradoxically, within a surprisingly short time you will find that the need to consult the key is rapidly diminishing.

This is not to imply that all keys are formidable; some of them just look that way. Here, the principal key (the "Pictorial Key") is a visual one, based upon comparing the appearance of an individual flower with a set of pictorial symbols. The "General Key" is merely an index to facilitate locating the appropriate chart in the Pictorial Key.

In practice, the General Key should be consulted first, after carefully noting the shape of the flower, to ascertain which of the 26 charts in the Pictorial Key is applicable. (Please note that in the General Key the term "perianth" is used to mean the showy or colorful part of the flower, and this may happen to be the corolla, or the calyx, or both; everywhere else in the book, however, "perianth" means the corolla and calyx collectively.)

Reference is then made to the appropriate chart in the Pictorial Key, where there may be a choice among several flower shapes. The last component to be factored in is the flower color, and this will direct you to a genus or family (sometimes more than one). All that then remains is to consult the descriptive text and corresponding photographs to determine the species.

It will soon become evident that an inexpensive hand lens of about 10-power magnification will be of help in examining small features. Wildflower identification can be carried out in the field with very little equipment, but this one item should be considered essential.

THE FINE PRINT

It is important to remember that the classification system and the keys we use in conjunction with it are artificial contrivances, created in an effort to re-cast what the human mind perceives as the chaotic diversity of nature into orderly patterns more suited to our powers of comprehension.

No key can possibly accommodate the myriad variations and subtle nuances that are so common in nature, short of incorporating into it all kinds of dis-tracting fine print in the form of footnotes, paren-thetical exceptions, and other qualifiers. To avoid such clutter, the following paragraphs point out some of the pitfalls that might be encountered as you work your way through the identification process.

For example, the General Key requires that you ascer-tain whether the "perianth" segments are really separate or are united at the base. The test is to pull one off to see if doing so disturbs the adjacent ones.

It also might be difficult to judge whether a peri-anth is regular or irregular; when in doubt, you will have no choice except to try both paths.

Charts U, V, and W represent a partial duplica-tion of taxa covered by Charts F through T, but pro-vide an alternate approach in cases where the indi-vidual flowers are minute.

Although the Composites (Asteraceae) might have been combined with others having individual flow-ers with either regular or irregular perianths, they have been separated because of their unusual mor-phology, and are traceable only through Charts X, Y, and Z.

The symbols provided in the Pictorial Key are in-tended to depict the floral structures schematically, not literally. Differences occur within almost every grouping; very few species look exactly alike.

The names of colors merely suggest the predomi-nant flower color, and cannot possibly encompass the many variations that will be seen in actual specimens. White-flowered individuals are to be expected in many species where colored flowers are the norm.

White flowers frequently are tinged with other colors, and yellow ones with green.

In the Pictorial Key, the names of families appear in roman type and the names of genera in italics.

In the main section of the book, which contains the descriptive text and photographs, the plants are arranged by families so that related genera may be considered together. This is especially important if you already know the family to which a plant belongs.

The first group of families, from the Typhaceae through the Orchidaceae, consists of the monocotyledons, which typically have parallel-veined leaves and 3- or 6-parted flowers. The remainder, which are the dicotyledons, are characterized by net-veined leaves and mostly 4- or 5-parted flowers. This subclass is itself divided according to whether the petals are separate (Saururaceae through Cornaceae) or united (Santalaceae through Asteraceae).

Within each family, the genera and species are arranged in more or less arbitrary sequence for ease of comparison, and simple supplementary keys are provided where genera and/or species are numerous. Scientific names are usually given first, with synonyms (names under which they may appear in other current publications) given in parentheses, followed by common names if any.

An important feature of the descriptions is the emphasis on those diagnostic characters that will be found most useful in distinguishing one species from another. It is advisable, therefore, to scan the descriptions of several apparently similar species, as well as the parts of the text that describe the genus and the family, in order not to miss any pertinent general information or mention of shared characteristics.

Because of the need for both accuracy and brevity, the use of technical terms could not be avoided, but definitions of these are provided in the comprehensive Glossary. A few of these terms are nouns used to designate the principal parts of plants; they are fundamental to the identification process and will soon become a part of your vocabulary. The others are mostly adjectives used to describe plant characteristics and, for the most part, can just as well be looked up as the need arises.

Individual plants within a given species often exhibit marked variations in shape, color, texture, and especially in dimensions. The text endeavors to describe "typical" examples as they would appear at the time of flowering and purposely avoids mention of extremes which might lead to unrealistic expectations.

If a species has a preference for an unusual habitat—such as a spruce-fir or cove hardwood forest, grassy or heath bald, or granite outcrop—this information is included. Because of wide fluctuations in flowering times, they are indicated in terms of seasons rather than months.

ADDITIONAL READING

There are available many excellent sources of information on the flora of the southern mountains, some of which are listed in the Bibliography.

The beginning wildflower enthusiast is encouraged to become familiar with such publications and to acquire the habit of cross-checking in cases of questionable identification. Most of the field guides listed provide only partial coverage of the southern mountains, but all can be very useful.

General Key ❧

See Pictorial Key Chart

Flowers without an evident perianth

Inflorescence cylindrical	A
Inflorescence club-shaped	B
Inflorescence ball-shaped	C
Inflorescence cup-shaped	D
Inflorescence of clustered stamens or pistils	E

Flowers regular

Perianth 2-parted	F
Perianth 3-parted or 3-lobed	G
Perianth 4-parted, segments separate	H
Perianth 4-lobed, segments united	I
Perianth 5-parted, segments separate	J
Perianth 5-lobed, segments united	K
Perianth 6-parted or 6-lobed	L
Perianth segments variable or numerous	M

Flowers slightly irregular

Perianth 4-parted or 4-lobed	N
Perianth 5-parted or 5-lobed	O

Flowers irregular

Perianth 3-parted or 3-lobed	P
Perianth 4-parted or 4-lobed	Q
Perianth 5-parted, segments separate	R
Perianth 5-lobed, segments united	S
Perianth 6-parted or 6-lobed	T

Small flowers in crowded inflorescences

Heads	U
Umbels or Cymes	V
Spikes or Racemes	W

Composites

Ray Flowers only	X
Disk Flowers only	Y
Ray and Disk Flowers	Z

Pictorial Key ༈

Chart	Flower Type	Color	Genus or Family	Page No.
A		Brown	*Typha*	1
B		Yellow Purple Green	Araceae *Arisaema* Araceae	3 3 3
C		Yellow Green	*Symplocarpus* *Sparganium*	3 1
D		White or Green	*Euphorbia*	92
E		White Yellow or Purple	*Cymophyllus* *Saururus* Ranunculaceae *Pachysandra* *Thalictrum*	2 30 43 93 50
F		White	*Circaea*	110
G		White Yellow	Alismataceae *Xyris*	2 4
		White Pink or Purple Blue	*Trillium* *Galium* *Murdannia* *Trillium* *Tradescantia*	15 163 5 15 5
		Brown	Aristolochiaceae	30
		Yellow or Purple	*Trillium*	15

Chart	Flower Type	Color	Genus or Family	Page No.
G (cont.)		Yellow or Blue	*Iris*	19
H		White	*Argemone*	54
			Brassicaceae	55
		Yellow	Papaveraceae	53
			Brassicaceae	55
			Hypericum	97
			Onagraceae	107
		Pink	*Rhexia*	107
			Epilobium	109
		Red	*Papaver*	53
		Purple	Brassicaceae	55
			Euonymus	94
			Rhexia	107
			Epilobium	109
		White	*Maianthemum*	11
			Clematis	45
			Sanguisorba	73
			Cornus	118
		Yellow	*Sedum*	63
			Hypericum	97
		Blue	*Clematis*	45
		Green	*Chrysosplenium*	68
		White or Pink	Onagraceae	107
		Yellow	*Oenothera*	107
		White	*Draba*	58
I		White	*Polygonum*	32
			Polypremum	123
			Frasera	125
			Rubiaceae	163
		Yellow	*Frasera*	125
			Galium	163
		Purple	*Galium*	163
		Blue	*Houstonia*	165
		White	Rubiaceae	163
			Veronicastrum	156
		Pink	Rubiaceae	163
			Dipsacus	167
		Blue	Rubiaceae	163
		Green	*Ludwigia*	108
		White or Blue	*Obolaria*	124

Chart	Flower Type	Color	Genus or Family	Page No.
I (cont.)		Blue	*Gentianopsis*	123
		White, Yellow, or Purple	*Clematis*	45
J		White	*Phytolacca*	35
			Ranunculaceae	43
			Drosera	63
			Saxifragaceae	64
			Rosaceae	68
			Hibiscus	95
			Araliaceae	110
			Apiaceae	111
			Chimaphila	118
		Yellow	Ranunculaceae	43
			Rosaceae	68
			Oxalis	88
			Linum	90
			Malvaceae	95
			Hypericum	97
			Helianthemum	99
			Ludwigia	108
			Apiaceae	111
		Orange	*Modiola*	96
		Pink	*Phytolacca*	35
			Portulacaceae	35
			Drosera	63
			Saxifragaceae	64
			Rosaceae	68
			Geranium	89
			Hibiscus	95
			Triadenum	99
			Chimaphila	118
		Purple	*Rubus*	72
			Geraniaceae	89
			Thaspium	117
		Purple or Green	*Euonymus*	94
		White	*Comandra*	30
			Mollugo	35
			Saxifragaceae	64
			Porteranthus	72
			Apiaceae	111
		White, Yellow, or Pink	*Sedum*	63
		Purple	*Xanthorhiza*	44
			Saxifragaceae	64
		Blue	*Eryngium*	112
		Green	Saxifragaceae	64
			Aralia	111

Chart	Flower Type	Color	Genus or Family	Page No.
J (cont.)		White	Caryophyllaceae	36
			Oxalis	88
			Malva	96
		Yellow	Portulaca	36
			Rosaceae	68
			Oxalis	88
			Abutilon	96
		Pink	Silene	37
			Geranium	89
			Malva	96
		Purple	Agrostemma	39
			Oxalis	88
			Geranium	89
		White	Caryophyllaceae	36
		Pink	Lychnis	37
		Red	Silene	37
		Pink	Dianthus	39
		White	Silene	37
		White	Pyrola	119
		Yellow	Nuphar	42
			Sarracenia	62
		Purple	Sarracenia	62
		White	Mitella	66
		White, Yellow, or Red	Monotropa	119
		Pink	Ericaceae	118
		Red	Aquilegia	44
K		White	Galax	120
			Samolus	122
			Phlox	132
			Boraginaceae	135
		Yellow	Lysimachia	121
			Lithospermum	137
			Verbascum	152
		Pink	Sabatia	125

Chart	Flower Type	Color	Genus or Family	Page No.
K (cont.)		Pink or Purple	*Phlox*	132
		Red or Blue	*Anagallis*	122
		Blue	*Vinca*	126
			Phacelia	135
			Boraginaceae	135
		White	Polygonaceae	32
			Menyanthes	125
			Cynanchum	130
			Solanum	151
			Sicyos	168
		Yellow	*Lysimachia*	121
			Solanum	151
		Pink	Polygonaceae	32
		Purple or Green	*Matelea*	129
		Blue	*Solanum*	151
			Campanulaceae	168
		Yellow	*Physalis*	149
		Blue	*Nicandra*	150
		White, Pink, or Blue	*Phlox*	132
		Yellow	*Melothria*	168
		Pink	*Mirabilis*	35
		White or Blue	*Phacelia*	135
		White	*Valerianella*	167
		Red	*Spigelia*	123
		Blue	*Amsonia*	126
			Valerianella	167
		White or Pink	*Apocynum*	125
		White or Blue	*Campanula*	168
		White	*Shortia*	120
			Cuscuta	131
			Hydrophyllum	134
		White or Pink	*Epigaea*	119
		Blue	*Polemonium*	134
			Hydrophyllum	134
		White	*Gaultheria*	119
		Blue	*Campanula*	168

Chart	Flower Type	Color	Genus or Family	Page No.
K (cont.)		White or Blue	Gentiana	123
		Yellow	Triosteum	166
		Pink	Schrankia	73
		Purple	Cynoglossum	137
			Triosteum	166
		Yellow	Onosmodium	137
		Blue	Gentianella	123
		Blue	Mertensia	136
		White	Datura	150
		White or Pink	Convolvulaceae	130
		Red	Ipomoea	130
		Blue	Ipomoea	130
			Ruellia	162
		White or Pink	Dodecatheon	122
		Various colors	Asclepias	127
L		White	Liliaceae	6
			Diphylleia	52
		Yellow	Smilax	17
		Pink	Helonias	13
		Blue	Camassia	13
		White	Liliaceae	6
			Sisyrinchium	20
			Echinocystis	168
		Yellow	Heteranthera	6
			Hypoxis	18
			Caulophyllum	52
		Orange	Belamcanda	21
		Purple	Brasenia	42
			Caulophyllum	52
			Lythrum	106
		Blue	Heteranthera	6
			Sisyrinchium	20
		Green	Liliaceae	6
			Dioscorea	19
		White	Melanthium	14

Chart	Flower Type	Color	Genus or Family	Page No.
L (cont.)		White	*Hymenocallis*	18
		White	*Yucca*	6
		Yellow	*Eriogonum*	34
			Nuphar	42
		White or Red	*Streptopus*	7
		Yellow	*Medeola*	11
		White	*Convallaria*	7
		Blue	*Muscari*	8
		White	Liliaceae	6
		Yellow	*Polygonatum*	7
		Blue	*Muscari*	8
		Green	*Dioscorea*	19
		White	Liliaceae	6
		Pink	*Allium*	11
		Yellow	*Uvularia*	8
			Manfreda	19
		White, Yellow, or Orange	Liliaceae	6
		Red	*Lilium*	10
		White	*Zephyranthes*	18
M		White	Ranunculaceae	43
			Berberidaceae	52
			Papaveraceae	53
		Yellow	*Caltha*	47
			Argemone	54
		Pink or Blue	*Hepatica*	44
		White	*Nymphaea*	42
			Trientalis	122
		Yellow	*Opuntia*	106
		Yellow or Purple	*Passiflora*	105

Chart	Flower Type	Color	Genus or Family	Page No.
N		White or Blue	*Veronica*	157
		White or Pink	*Phyla*	138
		Yellow or Purple	*Gratiola*	159
O		Various colors	*Viola*	100
		White or Purple	Apiaceae	111
			Verbena	138
			Pycnanthemum	141
		Blue	*Verbena*	138
			Scrophulariaceae	152
		White or Purple	*Orobanche*	162
		Yellow	*Aureolaria*	152
		Pink	*Agalinis*	153
		Red	Bignoniaceae	161
		White, Pink, Purple, or Blue	*Echium*	135
			Lamiaceae	139
P		Blue	*Commelina*	4
		White, Pink, or Purple	*Polygala*	90
		Yellow or Purple	*Aristolochia*	30
Q		White or Pink	*Gaura*	110
		White or Pink	Fumariaceae	54

Chart	Flower Type	Color	Genus or Family	Page No.
Q (cont.)		Yellow or Pink	*Corydalis*	55
		Yellow or Purple	*Epifagus*	161
R		Yellow	Caesalpiniaceae	74
		White or Blue	Ranunculaceae	43
		Green	*Hybanthus*	105
		Various colors	Fabaceae	75
		Yellow or Blue	*Aconitum*	43
		Yellow or Orange	*Impatiens*	94
S		Blue	*Justicia*	162
		White or Purple	*Lindernia*	160
		White, Pink, or Purple	*Penstemon*	153
		White, Blue, or Red Purple	*Lobelia* *Phryma*	169 162

Chart	Flower Type	Color	Genus or Family	Page No.
S (cont.)		Blue	*Mimulus*	155
		Yellow or Blue	Scrophulariaceae	152
		Various colors	Lamiaceae	139
		Yellow or Purple	*Collinsonia*	146
		Purple Blue	*Chaenorrhinum* *Mazus*	157 160
		Yellow Red Purple	Scrophulariaceae *Castilleja* *Pedicularis*	152 156 155
		Various colors	*Monarda*	139
		Pink	*Teucrium*	142
		White, Pink, or Purple Yellow	*Chelone* *Conopholis*	154 161
		Brown or Green	*Scrophularia*	156
T		Purple	*Cuphea*	106
		Blue	*Pontederia*	5

Chart	Flower Type	Color	Genus or Family	Page No.
T (cont.)		White or Purple	Orchidaceae	21
		Various colors	Orchidaceae	21
		Various colors	Orchidaceae	21
		Various colors	Orchidaceae	21
		Pink	*Calopogon*	24
		White, Purple, or Green	Orchidaceae	21
		White, Yellow, or Pink	*Cypripedium*	21
		Purple	*Galearis*	25
		Pink	*Arethusa*	22
		White	*Spiranthes*	28

Chart	Flower Type	Color	Genus or Family	Page No.
T (cont.)		White	*Goodyera*	29
U		White	*Trifolium*	84
			Eryngium	112
			Pycnanthemum	141
			Valerianella	167
		Yellow	Fabaceae	75
		Pink	*Schrankia*	73
			Fabaceae	75
			Phyla	138
			Dipsacus	167
		Red	*Trifolium*	84
		Purple	*Pycnanthemum*	141
		Blue	*Eryngium*	112
			Sherardia	165
			Valerianella	167
V		White	Liliaceae	6
			Araliaceae	110
			Apiacae	111
			Cuscuta	131
			Lycopus	144
			Galium	163
		Yellow	*Smilax*	17
			Apiaceae	111
			Galium	163
		Pink	*Allium*	11
		Purple	*Thaspium*	117
		Blue	*Hedeoma*	145
W		White	Liliaceae	6
			Orchidaceae	21
			Polygonum	32
			Brassicaceae	55
			Aruncus	73
			Fabaceae	75
			Polygala	90
			Galax	120
			Veronicastrum	156
		Yellow	*Rorippa*	57
			Agrimonia	69
			Melilotus	86
		Pink	*Helonias*	13
			Polygonum	32
			Heuchera	66
			Lespedeza	82
			Polygala	90

Chart	Flower Type	Color	Genus or Family	Page No.
W (cont.)		Purple	*Lespedeza*	82
			Polygala	90
			Lamiaceae	139
		Blue	*Muscari*	8
			Medicago	86
			Verbena	138
		Green	*Dioscorea*	19
			Malaxis	25
X			See Asteraceae (ligulate heads)	171
Y			See Asteraceae (discoid heads)	171
Z			See Asteraceae (radiate heads)	171

Descriptions of Species ❧

TYPHACEAE, CAT-TAIL FAMILY

Typha latifolia, Common Cat-tail (plate 1). The stiff, flat linear leaves of this Cat-tail are $^1/_2$ to 1 inch wide and 3 feet or more in length. An erect stalk nearly as tall bears a dense, velvety brown cylindrical spike 1 inch in diameter, composed of minute pistillate flowers; this is overtopped by and confluent with a smaller spike of staminate flowers that wither early. Shallow water and wet ditches. Summer.

Narrow-leaved Cat-tail, *T. angustifolia,* has leaves that are less than $^1/_2$ inch wide and convex on the back. The staminate and pistillate spikes are slender and are separated by a section of bare stalk.

Cat-tails not only grow abundantly in many parts of the world but for centuries have been valued as a food source. The young shoots that arise from the rhizome are peeled and cooked, and the copious golden pollen is used in place of flour.

SPARGANIACEAE, BUR-REED FAMILY

Sparganium americanum, Bur-reed (plate 2). Bur-reeds are frequently found along shorelines, in or near the water. The inflorescences arise from the axils of leaves that are dilated at their bases. The whitish or greenish unisexual flowers are borne in dense globose heads— staminate above and pistillate below. Spring–summer.

ALISMATACEAE, WATER PLANTAIN FAMILY

Alisma subcordatum, Water Plantain (plate 3). The long-stalked leaves of Water Plantain are elliptic to ovate, without basal lobes. Although the blades may reach 8 inches in length, the 3-petaled white flowers are less than $1/4$ inch across. They are borne in an open panicle with many whorled branches, and all are perfect, each having 6 to 9 stamens and a ring of pistils. Shallow water or mud. Spring–fall.

Sagittaria latifolia, Arrowhead (plate 4). This genus takes its name from the pointed lobes of the leaf bases, as typified by *S. latifolia,* our most common species (although unlobed leaves are also to be expected). Its flowers are in whorls of 3 or more—staminate ones above, and pistillate or perfect ones below on longer stalks—subtended by bracts that are shorter than the pedicels. The white petals are $1/2$ to $3/4$ inch long. Shallow water. Summer–fall.

Similar is **S. engelmanniana ssp. longirostra** *(S. australis),* which occurs less frequently in the mountains; it can be distinguished by its bracts, which are longer than the pedicels.

There are several Arrowheads with leaves that consistently lack the basal lobes, but only one approaches our region. It is Bunched Arrowhead, **S. fasciculata** (plate 5), a small plant (under 1 foot) with spatulate leaves and a few flowers measuring 1 inch across. It grows in shallow, slow-moving water and is endemic to the upper piedmont of the Carolinas. Spring.

CYPERACEAE, SEDGE FAMILY

Cymophyllus fraseri *(Carex fraseri),* Fraser's Sedge (plate 6). Were it not for its large size and showiness this species would have been excluded from this volume along with other grasslike plants. The cottony white flowers standing above a clump of strap-shaped leaves up to 2 inches wide are an arresting sight in the dark mountain woods, and its rarity makes it a worthy find. Spring.

ARACEAE, ARUM FAMILY

Arisaema triphyllum (A. atrorubens, A. stewardsonii), Jack-in-the-Pulpit, Indian Turnip (plate 7). This is a familiar wildflower, easily recognizable despite the variability of individual plants. Normally they have a pair of leaves 1 foot or more high and palmately divided into thirds. The solitary inflorescence consists of a club-shaped spadix with tiny flowers near its base, surrounded by a 2- to 4-inch vaselike spathe that terminates in a flap that arches over the top. The spathe varies from pale green to purplish brown, and is often striped. Moist woods. Spring–early summer.

Biting into a Jack-in-the-Pulpit plant causes an instant and painful burning sensation, due to numerous needlelike crystals of calcium oxalate, but Native Americans learned to eliminate this irritating property by baking and grinding the roots to make an edible flour.

Green Dragon, **A. dracontium,** is less well known but unmistakable. It has a very slender spadix 4 inches to much more in length, extending far beyond the end of the narrow spathe. There is only one leaf, which may have 7 or more leaflets. Wet woods. Spring–summer.

Peltandra virginica, Arrow Arum, Tuckahoe (plate 8). The foliage of this plant suggests that of *Sagittaria latifolia,* but its leaves are pinnately instead of palmately veined. The small yellowish flowers are borne on a slender spadix which is partially surrounded by the erect, pointed green spathe. In shallow water. Late spring–early summer.

Symplocarpus foetidus, Skunk Cabbage (plate 9). The thick, fleshy spathes of Skunk Cabbage appear in late winter or very early spring, sometimes emerging through a layer of snow. They are ovoid hoods tapering to a point at the summit, yellow-green to purple-brown, and often variegated. A yellowish 1-inch globose spadix covered with minute flowers is almost completely hidden within. The leaves develop later, eventually becoming very large. Swamps and wet woods.

Acorus calamus (A. americanus), Sweetflag, Calamus (plate 10). The inflorescence of Sweetflag is a 3-inch, elongate, tapered spadix completely covered by tiny brownish or greenish flowers. This projects at an angle from the middle of a stem, which resembles the sword-like leaves and extends above the spadix to form a linear spathe. Wet areas. Summer.

Orontium aquaticum, Golden Club (plate 11). The elliptic leaves of this aquatic are surpassed by white, finger-like scapes that terminate in a 2-inch spadix covered with bright yellow flowers. There is only a vestigial spathe at its base. The ability of the leaves to shed water has earned for it the popular name of "Never-wet." In slow-moving shallow water. Spring.

XYRIDACEAE, YELLOW-EYED GRASS FAMILY

Xyris torta, Yellow-eyed Grass (plate 12). The flowers of Yellow-eyed Grass are ephemeral and not likely to attract attention unless seen in numbers. This is a plant less than 20 inches high, with spirally twisted leaves only $^1/_{16}$ inch wide. A slender scape bears a conelike head covered with brown, scaly bracts, from which flowers emerge 1 or 2 at a time. Each has 3 bright yellow spreading petals, obovate and clawed. Wet places. Summer.

COMMELINACEAE, SPIDERWORT FAMILY

Commelina communis, Asiatic Dayflower (plate 13). Day-flowers are succulent plants with ephemeral flowers subtended by a folded, heart-shaped spathe. All have in common a pair of large upper petals which are always blue and a smaller lower petal which varies in color and size as to species. In this introduced species, the third petal is white and quite small, and the edges of the spathe are not joined. The leaves are ovate. Roadsides and waste areas. Spring until frost.

The native Slender Dayflower, **C. erecta,** is similar but has lighter blue upper petals and linear to lanceolate leaves, and the margins of the spathe are fused at the base. It prefers dry woods and rocky slopes.

Less frequently seen are **C. virginica,** an erect species in which the spathe is fused basally but the lower petal is blue and scarcely smaller than the upper ones; and the creeping **C. diffusa,** which has lanceolate leaves, an open spathe, and a very small blue lower petal.

The petals of Asiatic Dayflower—two showy and one insignificant—inspired Linnaeus to name the genus after the Commelijn family, a trio of seventeenth-century Dutch botanists; two attained prominence but the other never amounted to much.

Tradescantia subaspera, Zigzag Spiderwort. Spiderworts are tall (up to 3-foot) plants with long leaves that have dilated basal sheaths, and cymes of handsome flowers growing out of leaflike bracts. The flowers are radially symmetric, about 1 inch across, with 3 ovate blue-purple petals and 6 fertile stamens with bearded filaments. In this species, the blades of the principal leaves are wider than the basal sheaths when flattened. Woodlands, forest edges. Early summer.

Two other species have leaf blades that are narrower than the expanded sheaths and prefer drier habitats. **Tradescantia ohiensis** (plate 14) is a smooth, glaucous plant, with sepals that are hairless or have an occasional tuft. **T. hirsuticaulis** is pubescent, and the sepals bear both glandular and non-glandular hairs.

A special variety of Spiderwort has been developed which is capable of detecting toxic chemicals and radioactivity in the air, indicating their presence by changing the color of its petals and staminal hairs from blue to pink.

Murdannia keisak *(Aneilema keisak)* (plate 15). Superficially, *Murdannia* is suggestive of a miniature Spiderwort. It is, however, a decumbent plant with solitary (or a few) pink flowers growing from leaf axils. The petals are no more than 3/8 inch long. Wet shores. Late summer–fall.

PONTEDERIACEAE, PICKEREL WEED FAMILY

Pontederia cordata (includes *P. lanceolata*), Pickerel Weed, Wampee (plate 16). Pickerel Weed plants grow crowded together in shallow water, often forming vast colonies. The leaves are long-stalked, with cordate or lanceolate blades up to 8 inches long. Each plant produces a 5-inch spike of violet-blue flowers. They are funnelform and 2-lipped; each lobe is ³/₈ inch long and 3-parted, the upper bearing a pair of yellow spots. Three of the 6 stamens protrude conspicuously. Early summer–fall.

Heteranthera reniformis, Mud Plantain. Although they belong to the same family and inhabit shallow water or muddy shores, the plants in this genus are quite different from Pickerel Weed. Mud Plantain has floating kidney- or heart-shaped leaves up to 2 inches wide on long stalks, and emergent spikes of several light blue flowers with bilaterally symmetric corollas. These have a ¹/₄-inch tube and 6 slightly shorter linear lobes.

In Water Star Grass, **H. dubia,** the leaves are linear and about ¹/₈ inch wide, and often are submerged. It bears a solitary emergent, regular flower with a 1- to 1 ¹/₂-inch tube and linear yellow lobes about ³/₁₆ inch long. Both species bloom in summer and fall.

LILIACEAE, LILY FAMILY

The Lily Family is a very large one, and while there are many familiar flowers among its members, its great diversity makes it difficult to characterize as a unit. In general, however, they have these features in common: 1) a radially symmetric perianth consisting of 3 sepals and 3 similar petals; 2) a superior ovary; 3) six stamens; and 4) simple leaves with parallel veins. *Trillium* is a notable exception, with its net-veined leaves and dissimilar petals and sepals; another is *Maianthemum*, which has 4-merous flowers. Many genera have numerous small flowers in spikes, racemes, or panicles, but large flowers will be found in *Lilium* and *Hemerocallis*, and other types of inflorescences in the following:

> Umbels: *Clintonia, Medeola, Allium, Nothoscordum*
> Solitary on leafless stalks: *Erythronium*
> Solitary to few on leafy stems: *Streptopus, Polygonatum,*
> *Disporum, Uvularia*

Yucca filamentosa, Yucca, Adam's Needle, Spanish Bayonet (plate 17). This Yucca bears a panicle of creamy white, bell-like flowers on a stout stalk arising 5 feet or more from a basal rosette of stiff, sharp-pointed leaves. The word *filamentosa* refers to the frayed curls of threadlike fibers on the leaf margins. It is a southeastern native that frequently escapes from cultivation. Late spring–early summer.

Aletris farinosa, Colic-root, White Star Grass (plate 18), has a rosette of lanceolate leaves and a stem up to 1 foot tall with small scattered bracts and bearing a spikelike raceme of white flowers. The perianth is $^3/_8$ inch long, cylindrical with slightly flaring lobes; externally it is rough-textured, which gives it a coarsely granular appearance. Dry soil in open woods. Summer.

Convallaria montana (*C. majalis* var. *montana,* *C. majuscula*), Wild Lily-of-the-Valley (plate 19). Native to the United States and restricted to rich forests in some of our southern mountains, this is a close relative of **C. majalis,** the popular Lily-of-the-Valley which was introduced from Europe, is widely cultivated, and frequently escapes to the wild. *C. montana* is not colonial, however, and does not form dense stands. Also, its inflorescence is shorter, rising only to about half the height of the leaves. Both species have sheathing leaves and a one-sided raceme of fragrant, $^1/_4$-inch, bell-shaped white flowers in spring or early summer.

Streptopus roseus, Rose Twisted Stalk, Rose Mandarin (plate 20). The Twisted Stalks are 1 to 2 feet high, sometimes horizontally forked, with zigzag stems and alternate leaves. The flowers are bell-shaped (the tepals are $^3/_8$ inch long with recurved tips) and hang singly from a short pedicel beneath the axils. The leaves of *S. roseus* are sessile and the flowers are rose or purplish.

S. amplexifolius, known as White Mandarin, is similar but has clasping leaves and greenish white flowers. Both species are found in rich woods at high elevations in the South. Spring.

Polygonatum pubescens, Solomon's Seal (plate 21), has an unbranched, arching stem up to 3 feet long with alternate leaves that are soft-hairy on the veins beneath. It bears axillary pairs of $^1/_2$-inch, greenish yellow tubular flowers with short spreading lobes.

P. biflorum is similar, but the leaves are smooth and

the flowers greenish white. Very large plants with more flowers per peduncle are considered by some to be a separate species, *P. canaliculatum (P. commutatum)*. Woods. Spring–summer.

The name "Solomon's Seal" refers to the resemblance of the circular scars on the rhizome, left there by the withering of each successive year's flower stalk, to impressions on a wax seal.

Muscari atlanticum (M. *racemosum*), Blue Bottles. These popular spring flowers persist after cultivation and sometimes spread to open fields and waste places. The compact raceme is made up of blue cylindrical flowers under $^1/_4$ inch long, nodding on very short pedicels. The leaves are $^1/_8$ inch thick, linear, and nearly round in cross-section, drooping at the tip. Grape Hyacinth, *M. botryoides* is similar but has globose flowers and flat leaves up to $^3/_8$ inch wide.

Disporum lanuginosum, Yellow Mandarin, Fairy Bells (plate 22). Our species of *Disporum* are 1 to 2 feet high with nearly horizontal forking branches and alternate leaves. One to 3 flowers hang from the tips of these branches, where they tend to be obscured by the leaves. In *D. lanuginosum* the perianth consists of 6 narrow, greenish yellow segments with long flaring points.

Nodding Mandarin, *D. maculatum* (plate 23), has tepals that are wider and are white, spattered with tiny purple dots. Both species inhabit rich woods and bloom in spring.

Uvularia spp., Bellwort. The *Uvularias* are spring-blooming woodland plants with narrowly bell-shaped solitary or axillary flowers. In the first two species, the leaf bases encircle the stem and appear as though pierced by it.

Uvularia perfoliata, Perfoliate Bellwort (plate 24), has a solitary light yellow flower about 1 inch long. It can be distinguished by the orange granular glands on the inside of the tepals. Its leaves are smooth.

Large-flowered Bellwort, *U. grandiflora* (plate 25), has 1 to 3 flowers about 1 $^1/_2$ inches long; they are a deeper, golden yellow and smooth inside. The undersides of the leaves are white-downy. Both the tepals and the foliage are twisted, which gives the plant a droopy, wilted appearance.

Mountain Bellwort, **U. puberula** (*U. pudica*) is a slightly smaller, more delicate plant. Its leaves are sessile and have a lustrous sheen. There are 1 to 3 flowers, light yellow, $^3/_4$ to 1 inch long, with styles that are separate for half their length.

U. sessilifolia, known as Wild Oats, resembles *U. puberula* but differs in that it is colonial, the foliage is dull, and the flowers are solitary. Also, the styles are separate only in the upper third or fourth.

Erythronium americanum, Trout Lily, Adder's Tongue, Dogtooth Violet. By whatever name it may be known, this early spring wildflower heralds the season for many people. Each flowering plant has a pair of shiny, mottled leaves and bears a solitary 1- to 1 $^1/_2$-inch nodding lilylike bloom with a recurved, yellow perianth, with the sepals often purplish on the back. The long, protruding, rust-brown anthers are conspicuous, and the stigmas are not divergent. In moist woods, often forming large colonies.

The very similar **E. umbilicatum,** Trout Lily (plate 26), long ignored, is now thought to be the more common species in the South. It can best be distinguished when in fruit by the indented, rather than rounded, summit of the capsule.

White Trout Lily, **E. albidum,** which has a more western distribution, has a white perianth and spreading stigmas.

Clintonia borealis, Bluebead Lily (plate 27). Both of our species of *Clintonia* have broad, oval basal leaves with a shiny surface and a depressed midvein. C. *borealis* bears a loose cluster of up to 5 (occasionally more) $^3/_4$-inch slightly nodding, greenish yellow flowers. At high elevations in the South, especially in spruce-fir forests. Late spring.

Speckled Wood Lily, **C. umbellulata** (plate 28), is found in rich woods at lower altitudes. Its flowers have a spreading perianth, with lobes about ³/₈ inch long, white and more or less speckled with purple. They are disposed in a rounded umbel, and are more erect than those of *C. borealis*. Late spring.

Lilium philadelphicum, Wood Lily (plate 29). All five of our true Lilies have handsome flowers, all blooming in summer, and mostly whorled leaves. The Wood Lily is the only one with erect, rather than nodding, flowers and with tepals that are narrowed to a slender claw. The plant is under 3 feet; the orange-red flowers are 4 inches wide. Dry woods and openings.

L. superbum, Turk's Cap Lily. Our tallest Lily (8 feet) and the most abundantly flowered (20 or more on a single plant are not unusual). The tepals are yellow to orange, spotted, 3 inches long, and reflexed so that the stamens and pistil project conspicuously. A green stripe at the base of each segment forms a star inside the throat. The leaves are narrow, lanceolate, and numerous. Open woods and meadows. (These are frequently referred to as "Tiger Lilies," a name that properly belongs to a garden escape, *L. tigrinum*, which has numerous alternate, not whorled, leaves, some with black bulblets in the axils.)

L. michauxii, Carolina Lily (plate 30), also has flowers with strongly recurved tepals and exserted stamens (but no green star). It is a much shorter plant and usually bears fewer than 4 flowers. Its leaves are fewer and shorter than in *L. superbum*, and are distinctly widest above the middle. Woods.

L. grayi, Gray's Lily, Roan Lily (plate 31). This species may grow to 8 feet but has fewer flowers than *L. superbum*. They are about 2 ¹/₂ inches long, the segments only slightly flared, deep red on the outside with a yellow, purple-spotted throat. Meadows on high mountain balds.

The rare Gray's Lily is threatened not only by overgrazing, poaching, and land development but also by European wild boars, which apparently consider the bulbs a delicacy. These animals were brought into the Great Smoky Mountains region in 1912 for sport hunting, but eventually escaped from captivity and are uprooting vegetation on an alarming scale.

L. canadense, Wild Yellow Lily, Canada Lily, attains a height of 5 feet and its 3-inch flowers have gracefully wide-flaring tepals. The stamens scarcely project beyond the perianth. Flowers vary in color from yellow (the most common) to orange, and even brick-red in **ssp. editorum.** Wet meadows and coves.

Hemerocallis fulva, Orange Daylily (plate 32), is one of our more successfully naturalized garden flowers, thanks to its ability to proliferate by means of tuberous roots. Its 5-foot, leafless stalks produce several large (3- to 4-inch) flowers which open 1 or 2 at a time and last only for a day. The petals are slightly larger and have more ruffled margins than the sepals. Along roadsides and fields, often massed. Late spring–summer.

Medeola virginiana, Indian Cucumber Root (plate 33). Flowering plants of Indian Cucumber Root have two whorls of leaves—usually but not always 3 at the summit and 7 larger ones near the middle of the stem. The delicate, spidery flowers dangle on slender pedicels beneath the upper leaves and are easily over-looked. They have 6 greenish yellow tepals about $^1/_2$ inch long and strongly recurved, and 3 brown spreading, threadlike stigmas. Woods. Late spring.

Maianthemum canadense, Canada Mayflower, Wild Lily-of-the-Valley (plate 34). It is unfortunate that the name "Wild Lily-of-the-Valley" should have been applied to this plant, in view of the very different flower structure in *Convallaria.* In fact, those of *Maianthemum* are an anomaly in a family characterized by 3- or 6-merous flowers, in that they have 4 tepals and 4 stamens. It is short, seldom exceeding 6 inches in height, has 2 or 3 glossy leaves with clasping, heart-shaped bases, and bears a small raceme of $^3/_{16}$-inch white flowers. Moist woods, often forming dense mats. Spring–early summer.

Allium canadense, Wild Onion, Wild Garlic (plate 35). All of our species of *Allium* have the odor characteristic of onions. *A. canadense* has narrow, flat leaves and an erect stem bearing an umbel which contains a few white or pink flowers mixed with bulblets, or only bulblets, and is subtended by 2 or 3 bracts. Moist meadows or woods. Late spring–summer.

A. vineale, Field Garlic, is similar but with round, hollow leaves and a single bract. This is an all too familiar weed in lawns.

A. cernuum, Nodding Wild Onion (plate 36), has a many-flowered umbel without bulblets. A crook in the peduncle just below the umbel causes it to nod. The leaves are flat and less than $1/4$ inch wide, and the flowers vary from purple to pink to white. Rocky soil and open woods. Summer.

A. tricoccum, Wild Leek, also known as Ramps, blooms in summer after the flat, elliptic leaves (which are more than 1 inch wide and reddish at the base) have withered. Numerous white flowers are crowded in an umbel atop the naked stalk. Plants with fewer than 20 flowers and narrower leaves not tinged with red have been segregated as **A. burdickii,** White Ramps. Rich mountain woods.

Wild Leek bulbs are dug in the early spring, long before the flowers appear. Enthusiasts declare them to be one of the real gastronomic treats among the native plants of the southern Appalachians, and many communities stage ramp festivals annually. Dissenters, acutely aware of the clinging and persistent garliclike aroma, have a somewhat different opinion.

Nothoscordum bivalve (*Allium bivalve*), False Garlic (plate 37). This is similar to the *Alliums* but has slightly larger white flowers (the perianth segments are just under $1/2$ inch). The umbels are few-flowered, and the leaves are linear, $1/8$ inch wide. Dry woods, fields, and granite outcrops. Spring.

Ornithogalum umbellatum, Star-of-Bethlehem (plate 38). This handsome plant, a native of Europe long cultivated here, has escaped in many places to roadsides, meadows, and low woods. Its grasslike leaves have a white midrib, and this is echoed in a broad green stripe on the underside of each white tepal. The flowers, which open only in bright sunshine, are $1 1/4$ inches across and are borne in a corymb. Spring.

Camassia scilloides, Wild Hyacinth (plate 39), has flowers of light blue—an unusual color for the Lily Family—in a bracted raceme. There are 6 similar perianth parts, spreading and narrowed to a claw at the base. Its leaves are grasslike and basal. Moist places. Late spring–early summer.

A similar species (C. quamash), commonly called simply "Camass," grows in great numbers in the northwestern United States. Its sweet, edible bulbs were an extremely important food for both Indians and settlers, and probably saved the Lewis and Clark expedition from starvation. Unfortunately, its common name has been misapplied to poisonous plants in the genus Zigadenus (as in "Death Camass"), with sometimes tragic consequences.

Helonias bullata, Swamp Pink (plate 40). This is one of the most beautiful of all the plants that inhabit cool swamps and bogs. A 2-inch, egg-shaped cluster of pink flowers with pale blue anthers, and exuding a spicy fragrance, is borne on a hollow stalk above a rosette of oblanceolate evergreen leaves. Spring.

Smilacina racemosa (*Maianthemum racemosum*), False Solomon's Seal (plate 41). Vegetatively, this resembles the true Solomon's Seals (*Polygonatum*) but its flowers have narrow, white, spreading tepals and are in a 4-inch terminal panicle. Woods and forest edges. Spring. **S. stellata** (*Maianthemum stellatum*), Starry Solomon's Seal (plate 42), is erect with narrow, ascending leaves and a short raceme of $^3/_8$-inch flowers with spreading tepals. Moist sandy soil.

Xerophyllum asphodeloides, Turkey Beard (plate 43). The large basal tuft of long, wiry evergreen leaves only $^1/_{16}$ inch wide reveals the kinship of Turkey Beard with X. *tenax*, the Beargrass of the Northwest. The 4-foot, rigid, unbranched stem is invested with many shorter, stiffly ascending leaves and is surmounted by a dense raceme of white, $^3/_8$-inch flowers on 1 $^1/_2$-inch pedicels. Dry open woods. Late spring–early summer.

Amianthium muscaetoxicum, Fly Poison (plate 44). With its compact terminal raceme of white flowers on long pedicels, Fly Poison might seem to resemble Turkey

Beard, but the leaves of Fly Poison are linear and up to
$3/4$ inch wide, and the stem leaves are lanceolate and
spreading. The perianth turns green after pollination
and remains so for some time before wilting. Woods
and open grassy areas. Early summer.

Chamaelirium luteum, Fairy Wand, Devil's Bit, Blazing Star (plate 45). *Chamaelirium* is dioecious—that is,
the staminate and pistillate flowers occur on separate
plants. The dense racemes of white flowers appear spikelike owing to the short pedicels; the staminate ones taper
to a drooping tip, while the pistillate ones are more
cylindrical and shorter at first but later elongate. The
basal leaves are obovate, up to 6 inches long, and the
stem has several greatly reduced linear leaves. Rich
woods. Spring.

Stenanthium gramineum, Featherbells (plate 46), is a
tall plant with a leafy stem and a large panicle of white
flowers terminating in a wandlike raceme, its lower
branches often curving downward. The perianth segments are $1/4$ inch long, narrow, and acuminate. Woods,
thickets, meadows. Summer–fall.

Melanthium latifolium (M. *hybridum*), Bunchflower
(plate 47). Bunchflowers are 3 feet tall or more, with
panicles of greenish white flowers that darken as they
age. The perianth segments are narrowed to a claw,
and each bears a pair of glands at the base of the
blade. In this species the tepals are ovate, up to $1/2$
inch long, and the margins are conspicuously wavy
and crisped. Its leaves are oblanceolate, 1 to 2 inches
across at the widest point. Rich woods and meadows. Summer.

A second species, **M. *virginicum*,** has linear leaves less
than 1 inch wide. Its flowers are slightly smaller, and
the margins of the tepals are entire and flat. Wet woods
and meadows. Summer.

M. *parviflorum* (*Veratrum parviflorum*). More than just
tinged with green, as in the last two, this plant and the
next have flowers that are distinctly yellow-green
throughout. Those of M. *parviflorum* are $1/2$ inch across
with smooth perianth segments that taper gradually to

the base; they are not clawed and have no glands. Its leaves are mostly near the base, elliptic, long-stalked, and under 8 inches long. Rich woods. Summer–fall.

Less widespread in the South is **Veratrum viride,** False Hellebore (plate 48), also known as Indian Poke. It has heavily ribbed, oval leaves that are up to 12 inches long and sessile or somewhat clasping and continue on the stem up to the inflorescence. Its flowers are slightly larger than in the preceding species, and have ciliate margins. Wet woods, swamps. Late spring–summer.

Zigadenus glaucus, White Camass (plate 49). Our plants in this genus have linear leaves, mostly crowded near the base, and a panicle of white or greenish flowers. This species has a purplish stem that is whitened with a waxy bloom. The perianth segments are about 3/8 inch long and each has a green bilobed gland near the base. Moist soil. Summer.

Z. leimanthoides is not glaucous and has a more densely flowered pyramidal inflorescence. The tepals are only $^3/_{16}$ inch long, and have a single obscure gland. Wet woods. Summer.

Tofieldia glutinosa (*T. racemosa* var. *glutinosa*), False Asphodel (plate 50). This is primarily a northern bog plant but is occasionally found in wet situations in the southern mountains. It has several 2-ranked narrow leaves near the base and a sticky glandular stalk bearing one small leaf near the middle. The flowers are white with $^3/_{16}$-inch tepals, and are borne in small clusters in a terminal raceme. Summer.

Trillium spp., Trillium, Wake Robin, Toadshade. Several features make the *Trilliums* easily recognizable— among them, a single whorl of three leaves, netted (instead of parallel) veins, a solitary terminal flower, and three green sepals that are very different from the three larger white or colored petals. All are woodland plants, blooming in the spring.

For convenience, the *Trilliums* are divided here into two groups: 1) those in which the flowers are stalked and face outward or downward and have spreading petals;

and 2) those with sessile flowers in which the petals stand erect, and with leaves that are often mottled.

> Flowers stalked: *T. grandiflorum*, *T. undulatum*,
> *T. erectum*, *T. simile*, *T. vaseyi*, *T. cernuum*, *T. catesbaei*,
> *T. pusillum*, *T. persistens*
> Flowers sessile: *T. luteum*, *T. cuneatum*, *T. decumbens*,
> *T. discolor*

Flowers stalked

Trillium grandiflorum, Large-flowered Trillium, White Wake Robin (plate 51). This is perhaps the best-known species. Its flowers are 2 to 3 inches long, funnel-shaped with the petals flaring widely and ruffled, white but quickly fading to pink.

T. undulatum, Painted Trillium (plate 52). Dainty in appearance but preferring cold forests at high altitudes, this Trillium has spreading, slightly recurved petals 1 ½ inches long, wavy-edged, white with crimson veining that forms a "V" at the base. Its foliage has a bronze tinge.

T. erectum, Purple or Red Trillium, Wake Robin, Stinking Benjamin (plate 53 and plate 54), has flowers that are held erect on a rigid stalk and face forward, the 1 ½-inch lanceolate petals spreading from the base to expose the brownish maroon ovary. Petal color varies from dark purplish red through pink and greenish tints to white, but all have a dark ovary. (Some plants with broad, ovate white petals are considered a separate species, **T. simile.**)

T. vaseyi, Vasey's Trillium (plate 55), formerly regarded as a variety of *T. erectum*, has a large maroon-purple flower that declines below the leaves and is characterized by ovate, recurved petals and conspicuous long, protruding anthers. It is our largest species.

T. cernuum, Nodding Trillium (plate 56), also has wide recurved petals and a flower that hangs below the leaves, but it is white or pale pink with a purplish ovary and anthers.

T. catesbaei (*T. nervosum*), Catesby's Trillium (plate 57), has a tendency to nod but not always beneath the leaves. Its flowers are pink or white with 1 ½-inch wavy-margined and recurved petals, narrow sickle-shaped sepals, and bright yellow anthers that are curved at maturity.

Much more scarce in our region, and bearing smaller flowers, is Least Trillium, **T. pusillum.** Its flowers are held erect and have narrow petals less than 1 inch long, white but turning to pink with age, and straight yellow anthers; the sepals are about as long as the petals. Its leaves are narrowly oblong and obtuse.

Another small-flowered species is **T. persistens.** Its erect white flowers open early in the spring and age to dark rose. The ovate leaves taper gradually to a point. Very rare, it has been found only in northeastern Georgia and nearby South Carolina.

Trillium persistens *is one of a very small number of plants that were not discovered—or at least recognized as being new to science—until after the middle of the twentieth century. The fact that such events can still occur is an indication that the southern mountain flora may be even richer than previously supposed.*

Flowers sessile

T. luteum (*T. cuneatum* var. *luteum*), Yellow Toadshade (plate 58), is a particularly handsome example of this group, with its distinctly yellow lanceolate petals 2 $^1/_2$ inches long and a greenish ovary.

Very similar is **T. cuneatum** (*T. hugeri*), known as Little Sweet Betsy (plate 59). Its petals are usually maroon or purple, but may vary to yellowish bronze. The ovary is purple.

Trailing Trillium, **T. decumbens,** another species with purplish petals, is unique in that its stem is decumbent, allowing the leaves to lie on the ground. One of the earliest to bloom, it is limited to the southern part of our range.

Pale Yellow Trillium, **T. discolor,** is a smaller plant than any of these, with leaves no more than 2 $^1/_2$ inches long and pale yellow spatulate petals 1 $^1/_2$ inches long.

Smilax herbacea, Carrion Flower (plate 60). As must be expected of flowers that are pollinated by flies, those of Carrion Flowers are ill-smelling. Unlike the woody species of *Smilax*, they lack thorns. *S. herbacea* is a vine, climbing by tendrils. The leaves are ovate with truncate or rounded bases, and the flowers—staminate and pistillate on separate plants—are greenish yellow, in rounded axillary umbels. Less common is **S. pseudochina** (*S. tamnifolia*), with cordate-based hastate leaves. Late spring–early summer.

S. biltmoreana (*S. ecirrhata* var. *biltmoreana*) grows erect, without tendrils, its leaves concentrated in the upper part of the plant. The umbels are on long axillary peduncles.

AMARYLLIDACEAE, AMARYLLIS FAMILY

This family is very closely related to the Liliaceae, differing mainly in its inferior ovary, and, because this distinction is sometimes blurred, the two are occasionally combined.

Hypoxis hirsuta, Yellow Star Grass (plate 61). This engaging little flower seems to put in an appearance during every woodland walk from spring until fall. Its ³/₄-inch flowers of 6 similar tepals are carried just a few inches above the ground and are overtopped by the hairy, grasslike leaves.

Zephyranthes atamasco, Atamasco, Zephyr, or Rain Lily (plate 62). A colony of these lilylike flowers in a wet woodland or grassy clearing is a most impressive sight. The perianth segments are fused into a 3- to 4-inch funnel-shaped flower which stands erect above foot-long grasslike leaves. They are white at first but fade to pale pink. This is essentially a lowland species, but a town in the mountains of western North Carolina was given the Native American name for the flower, *Cullowhee*, because of its abundance there. Spring.

Hymenocallis caroliniana (*H. occidentalis*), Spider Lily (plate 63). Several large flowers are borne in an umbel on a stalk rising 18 inches or more above a rosette of linear leaves. The tepals are white, very narrow, and 2 to 4 inches long, and spread widely. The lower portions of the long filaments are united by a funnel-shaped white "corona" with large teeth on its margin. Moist woods and stream banks. Summer.

Manfreda virginica (*Agave virginica*), False Aloe. This relative of the Century Plants has thick, succulent, sharp-pointed leaves in a basal rosette. There is a loosely flowered spike up to 6 feet tall; the flowers are greenish yellow, tubular at the base, and toothed at the apex, about 1 inch long, with exserted stamens. Dry, rocky woods. Spring–summer.

DIOSCOREACEAE, YAM FAMILY

Dioscorea villosa (*D. quaternata*), Wild Yam (plate 64), is a high-climbing, twining vine. The 2- to 4-inch leaves have heart-shaped bases and convex sides, are abruptly tapered to a long point, and have conspicuous parallel veins connected by numerous cross veins. The flowers are small and yellow-green, the staminate ones in axillary panicles, the pistillate ones on other plants in axillary spikes. Open woods and thickets. Spring–early summer.

Cinnamon Vine, **D. batatas,** is similar but has hastate leaves with concave sides. It frequently bears small gray tubers in the axils. Alluvial woods, waste places. Summer.

IRIDACEAE, IRIS FAMILY

Iris cristata, Crested Dwarf Iris (plate 65). The blue flowers of the Dwarf Irises are usually no more than 5 inches above the ground. This species is distinguished by a crinkled yellow crest on the sepals. Its leaves are usually more than 3/8 inches wide. Rich woods. Spring.

I. verna var. smalliana, Dwarf Iris (plate 66). This resembles *I. cristata*, but the flowers are more violet than blue, and the sepals bear a smooth orange-and-white blaze without a raised crest. The leaves are more erect and narrower. Dry woods. Spring.

I. virginica, Blue Flag, is the only tall (up to 3 feet) native Iris usually found in the southern mountains. Its flowers are about 3 inches across, light blue-violet with darker veins; the sepals have a yellow spot near the base. Stream banks and shallow water. Spring.

I. pseudacorus, Yellow Flag, Water Flag (plate 67). This is an alien species (we have no native yellow Iris) commonly escaping to streamsides. The flowers are 3 to 4 inches across. Late spring–summer.

The design of the device we call "Fleur-de-lis" is based not on a Lily but on the floral structure of an Iris. The name is a corruption of "Fleur de Louis," which was given to it when it first appeared on the banner of Louis VII during the Crusades.

Sisyrinchium **spp.,** Blue-eyed Grass. Differentiating between species in this genus is rather difficult, due partly to wide disagreement about the nomenclature. All have tufts of grasslike leaves and flattened, winged stems (scapes) that closely resemble the leaves. Each scape has one or more bracted spathes that bear clusters of flowers on threadlike stalks. The perianth has 6 similar segments (usually blue) and is about ³/₄ inch across. They may be found in woods or open places and start to bloom in spring, sometimes continuing until fall.

Two of our species have simple, unbranched stems terminating in a solitary sessile spathe. *Sisyrinchium mucronatum* has very narrow leaves and scapes, the latter only ¹/₁₆ inch wide, and its spathes are purple. *S. montanum* is similar but has wider leaves and scapes.

A third species with simple stems, *S. albidum,* has a twinned pair of spathes at the top of the scape, subtended by a single large outer bract. Its flowers may be either white, as the specific epithet suggests, or blue.

In *S. angustifolium* (plate 68) the stems are usually branched, each with 2 (occasionally more) spathes on long peduncles. The scapes are conspicuously winged, between ¹/₈ and ¹/₄ inch wide, and about equal in length to the leaves.

S. atlanticum is another branched species with 2 long-stalked spathes, but its scapes are less than ¹/₈ inch wide and much longer than the leaves.

S. dichotomum, White Irisette, is a federally endangered North and South Carolina endemic. It has 3 or more nodes and several pedunculate spathes that become progressively smaller upward.

Belamcanda chinensis, Blackberry Lily (plate 69). Except for having alternate sword-shaped leaves, the exotic Blackberry Lily is very unlike our other representatives of the Iris Family. It has orange flowers copiously spotted with crimson (the perianth segments are nearly all alike), less than 2 inches across, in a branched inflorescence. The fruit is a cluster of shiny seeds resembling a blackberry. Frequently escaped from gardens, in dry, open situations. Summer.

ORCHIDACEAE, ORCHID FAMILY

There are more kinds of flowering plants in this family than in any other, the great majority of them epiphytic— that is, attached to other plants but not deriving nutrients from them—and growing in the tropics. Ours are all terrestrial, and for the most part they are less flamboyant although many are admirable for their form and color.

Orchid flowers are 3-merous but zygomorphic; one petal (usually the lower) is markedly different from the pair of lateral ones and is referred to as the lip. They also are unique in that the stamens are united with the style to form a central structure called the column.

The genera can be roughly divided into three groups to facilitate identification:

> Flowers 1 to 3: *Cypripedium, Arethusa, Pogonia, Cleistes, Triphora, Isotria*
> Flowers more than 3; leaves absent at anthesis: *Aplectrum, Tipularia, Corallorhiza, Hexalectris, Spiranthes* (part)
> Flowers more than 3; leaves present: *Calopogon, Galearis, Listera, Liparis, Malaxis, Platanthera, Coeloglossum, Spiranthes* (part), *Goodyera*

Cypripedium acaule, Pink Lady's Slipper, Moccasin Flower (plate 70). Our *Cypripediums* have broad leaves and flowers with large, inflated, pouchlike lips and spreading lateral petals. The upper sepal arches over the flower and the lower pair are fused beneath. *C. acaule* has 2 basal leaves and a scape bearing a solitary flower.

The drooping lip is 1 to 2 inches long, deeply cleft on the upper surface, normally rose with darker veins but varying to greenish white. Woods. Spring.

C. calceolus var. pubescens, Yellow Lady's Slipper (plate 71), has several cauline leaves and may have 1 or 2 flowers on a stem. The yellow lip is from 1 $^{1}/_{4}$ to 2 inches long, and the lateral petals brownish or greenish yellow and twisted. In **var. parviflorum** the lip is less than 1 inch long, and the petals are purplish brown, proportionately longer, and tightly twisted. Rich woods. Spring.

C. reginae, Showy Lady's Slipper (plate 72), is a larger, more northern species. It has a leafy stem and 1 to 3 flowers; the lip is about 2 inches long, furrowed, rose-purple streaked with white. The sepals and lateral petals are white, the latter not twisted. Swamps and wet woods. Early summer.

The structure of Cypripedium *flowers helps to diminish the chances of self-pollination by forcing nectar-seeking insects to follow a one-way path through the pouch. Entering through the obvious front opening, they find they cannot reverse direction but must take the rear exit, which is designed to wipe off any previously collected pollen onto the stigma and deposit fresh pollen on their bodies for transfer to the next flower.*

Arethusa bulbosa, Bog Rose, Dragon's Mouth (plate 73). This small plant, which seldom exceeds 10 inches, has a single grasslike leaf that appears after flowering. The sepals and lateral petals of its solitary flower, all magenta-pink and similar in shape, are erect or slightly arching, while the lip curves sharply downward; it is whitish marked with purple and has a yellow-fringed ridge down the center.

Pogonia ophioglossoides, Rose Pogonia, Snake Mouth Orchid (plate 74), is a delicate plant up to 1 foot high, usually bearing a solitary pink to white flower. The narrow sepals are spreading, the petals broader and converging above the lip, all about $^{3}/_{4}$ inch long. The spatulate lip is horizontal, deeply fringed, with red veins and median rows of greenish yellow bristles. There is a single clasping leaf halfway up the stem and a smaller leaflike bract beneath the flower. Open, wet areas. Summer.

Cleistes divaricata, Rosebud Orchid, Spreading Pogonia (plate 75), has all of its petals joined to form a pink 2-inch tubular lip. The linear brownish or purplish sepals are even longer and spread widely above the lip. As in *Pogonia*, there is a solitary cauline leaf and a floral bract. Open woods and moist meadows. Late spring–summer.

Triphora trianthophora, Three Birds Orchid, Nodding Pogonia (plate 76), is a small, succulent plant with $^1/_2$- to $^3/_4$-inch ovate leaves that clasp the stem. Nodding on a stalk from each of the uppermost axils is a $^3/_4$-inch pale pink or white ephemeral flower. The sepals and lateral petals are similar and more or less spreading; the lip has 3 green ridges. Rich woods. Summer–early fall.

Isotria verticillata, Whorled Pogonia (plate 77), may grow to 16 inches tall and has a whorl of 5 or 6 oblong leaves, 3 inches or longer. Above this is a solitary long-stalked terminal flower with a $^1/_2$-inch lip, white or yellowish marked with purple. The sepals are linear and brown-purple, 2 inches long, the median one erect and the lateral pair extending forward. Rich woods. Spring–summer.

I. medeoloides, Small Whorled Pogonia (plate 78), an extremely rare species, has glaucous pale green, drooping leaves, a short flower-stalk, and sepals less than $^3/_4$ inch long. Occasionally there are two flowers.

Aplectrum hyemale, Puttyroot, Adam-and-Eve (plate 79). In late summer the corms of this plant produce a solitary elliptic leaf up to 6 inches long, grayish green with many narrow, white longitudinal ridges; this persists throughout the winter but withers before the naked flowering stalk develops. The flowers are in a raceme, about $^1/_2$ inch long, varying from yellowish to greenish and tinged with purple; the lip is white marked with magenta, not spurred. Rich woods. Late spring–early summer.

The fact that Aplectrum hyemale *produces a pair of corms led to the name "Adam-and-Eve." These enlarged underground stems had some early medical applications, but perhaps their oddest use was in the preparation of a mending cement for crockery, which earned for it another name, "Puttyroot."*

Tipularia discolor, Cranefly Orchid (plate 80), also forms a single over-wintering leaf that is absent at time of flowering, but it is dark green above and glossy purple beneath. There are numerous flowers in a raceme; they are pale greenish purple with narrow, widely spreading sepals and petals and an extremely slender spur $^3/_4$ inch long. Woods. Summer.

Corallorhiza trifida, Northern Coralroot. These leafless saprophytes take their name from their branched, coral-like rhizomes. The flowers are small, in a loose raceme. *C. trifida* has a slender, pale green stem and yellowish flowers with a $^1/_8$-inch white lip bearing 2 short lateral lobes. Wet woods. Spring–early summer.

C. maculata, Spotted Coralroot, may be reddish, purplish, brownish or yellowish. The lip is $^3/_8$ inch long, white with irregular magenta spots, and has a pair of prominent lateral lobes. Woods. Summer.

C. wisteriana, Spring Coralroot (plate 81), has yellowish or greenish flowers diffused with purple or brown. The lip is oval, entire, and white spotted with pink. Woods. Spring.

C. odontorhiza, Late Coralroot. In this species the stem has a thickened, bulbous base. The flowers, which open only slightly, are small and purplish green with a purple-spotted white lip that is nearly circular and has wavy margins. Woods. Fall.

Hexalectris spicata, Crested Coralroot. This resembles *Corallorhiza* but is larger in all respects; the scape may be 30 inches tall or more. The lateral sepals and petals are $^3/_4$ inch long, yellowish with purplish brown stripes; the lip is white with purple veins and several fleshy ridges. Woods, usually in calcareous soil. Summer.

Calopogon tuberosus (*C. pulchellus*), Grass Pink (plate 82). The most unusual feature of this genus is that the yellow-bearded lip is uppermost. The other petals and sepals are alike in shape and color—usually magenta-pink. In this species the entire flower is 1 inch or more wide, and there are several in a loose raceme. A solitary grasslike leaf sheathes the stem near its base. Wet places. Spring–summer.

Galearis spectabilis *(Orchis spectabilis)*, Showy Orchis (plate 83). This is a low plant with 2 lustrous, wide, elliptic leaves at the base. There are several 1-inch flowers in a short raceme; the lilac-colored sepals and lateral petals are combined to form a rounded hood, and the lip is white, nearly flat, points downward, and is spurred. Rich woods. Spring–early summer.

Listera smallii, Appalachian or Kidney-leaved Twayblade (plate 84). This small summer-flowering orchid grows in damp woods, almost always under Rhododendrons. There is a single pair of sessile ovate leaves midway between the base of the stem and the few-flowered raceme. The flowers are dull yellowish or greenish with a prominent, wedge-shaped lip; there are two small lateral lobes near its base, and it is deeply cleft at the broad apex with a minute tooth in the sinus.

L. australis, Southern Twayblade, is another small orchid with a pair of slightly larger ovate leaves on the stem. Its reddish purple flowers have a linear lip that is divided almost to the base. Wet woods. Spring–summer.

Liparis lilifolia, Lily-leaved Twayblade, Mauve Sleekwort (plate 85). In this genus the 2 leaves are much larger and basal. *L. lilifolia* has an obovate, translucent, madder-purple lip $3/8$ inch long and nearly as wide, with a tooth at the apex; the other petals are threadlike, and the sepals are greenish and slender. Rich woods. Spring–summer.

In **L. loeselii,** known as Fen Orchid or Bog Twayblade because of its preference for wetter habitats, the flowers are greenish yellow and the perianth parts less than $1/4$ inch long.

Malaxis unifolia, Green Adder's Mouth (plate 86). Like the Twayblades, this is a small, delicate plant, but it has only one leaf—which is relatively large, oval, shiny, and clasping the stem. The green flowers

are tiny, the whitish lip less than ¹/₈ inch long with 2
pointed lobes and a minute intermediate tooth. They
are numerous, crowded in a cylindrical spike. Moist,
open woods and grassy places. Summer.

Platanthera spp. Long known as Rein Orchids, this
large group comprises plants with racemes of flowers
varying in size, form, and color, and blooming for
the most part in summer. The characteristics of the
lip, with its basal spur, are often diagnostic.

> Flowers purple: *P. grandiflora, P. psycodes, P. peramoena*
> Flowers orange: *P. ciliaris, P. cristata*
> Flowers white: *P. blephariglottis, P. integrilabia*
> Flowers yellowish or greenish: *P. lacera, P. orbiculata,*
>   *P. clavellata, P. flava*

Platanthera grandiflora (*Habenaria psycodes* var.
grandiflora, H . fimbriata), Large Purple Fringed Orchid (plate 87). This is a handsome species with lilac or rose-purple flowers in a cylindrical raceme at least
2 inches thick. The lip is ³/₄ inch or more across and is
divided into 3 fan-shaped lobes which are deeply
fringed. The spur is slender and about 1 inch long. Wet
meadows and wood margins.

P. psycodes (*Habenaria psycodes*), Small Purple Fringed
Orchid, differs mostly in its dimensions. The inflorescence is usually 1 ¹/₂ inch or less in diameter, the lip ¹/₂
inch wide, and the spur less than 1 inch long. The segments of the lip are less deeply fringed.

P. peramoena (*Habenaria peramoena*), Purple Fringeless
Orchid. Much less common than either of the preceding
species, *P. peramoena* has a 3-parted lip with spatulate segments that are not fringed but may have erose margins;
the median lobe has a distinct notch at its apex. Wet places.

P. ciliaris (*Habenaria ciliaris*) (plate 88). Although this
is called Yellow Fringed Orchid, the color of its flowers
is apricot-orange. The cylindrical raceme is usually
densely flowered and about 2 inches in diameter. The
lip is oblong and unlobed but deeply fringed; it is ¹/₂
inch long, half the size of the slender spur. Woodland
margins, meadows, and bogs.

P. cristata (*Habenaria cristata*), Crested Fringed Orchid. This species is rather similar but its inflorescence is only about 1 ¹/₄ inch in diameter. The individual flowers are one-half as large as those of *P. ciliaris*, and the spur is shorter than the ovate lip.

P. blephariglottis (*Habenaria blephariglottis*), Large White Fringed Orchid (plate 89). In many respects the description of the Yellow Fringed Orchid would fit this species, the most striking difference being the pure white or, at the most, creamy-tinged flowers. The slender, gracefully curving spur may attain a length of 2 inches. Wet meadows and bogs.

P. integrilabia (*Habenaria blephariglottis* var. *integrilabia*). Known as Monkey Face, this southern species has a spatulate rather than oblong lip, and, instead of being fringed, it may have either entire or finely toothed margins.

P. lacera (*Habenaria lacera*), Ragged Fringed Orchid (plate 90). In this species the lip is deeply 3-lobed into wedge-shaped segments. The terminal lobe is long-clawed and deeply fringed at the end only, but the lateral ones are incised nearly to the base; all terminate in hairlike filaments. The spur is slender, nearly 1 inch long. The color of the flowers varies from pale yellowish to whitish green. Wet meadows, thickets, and bogs.

P. orbiculata (*Habenaria orbiculata*), Large Round-leaved Orchid (plate 91). This is the only one of our *Platantheras* that is normally found in dark forest shade. It is also unusual for its principal foliage, which consists of 2 large, nearly circular basal leaves that lie flat on the ground (instead of the erect lanceolate to linear leaves of all the preceding species). The inflorescence is a loose raceme of greenish white flowers with a downward pointing, unlobed, linear-oblong lip ³/₄ inch long and a spur approximately twice as long.

P. clavellata (*Habenaria clavellata*), Small Green Wood Orchid (plate 92). This is a common species, usually less than 1 foot in height, with a solitary lanceolate clasping leaf and a few small bracts below a short, open raceme. The flowers are yellowish or greenish white, the lip less than ¹/₄ inch long, broadly wedge-shaped, and bearing 3 blunt teeth at the apex. The spur is longer, bulbous at the tip, curved, and turned to one side. Swampy woods, bogs, and wet, grassy road banks.

P. flava var. herbiola (*Habenaria flava* var. *herbiola*), Tubercled Rein Orchid. This has several lanceolate leaves and a compact raceme of yellowish green flowers interspersed with narrow bracts which extend beyond the flowers. The lip is less than ¹/₄ inch long with an irregular margin and a small protuberance near the base; the spur is somewhat longer. Wet woods.

Coeloglossum viride var. virescens (*Habenaria viridis* var. *bracteata*), Long-bracted Orchid. This resembles some of our *Platantheras* but differs in having an inconspicuous pouchlike spur that is smaller than the lip. It is a rather tall plant, up to 2 feet, with several clasping leaves gradually reduced above and passing into linear bracts which exceed the flowers. The flowers themselves are green, often tinged with purple, and have a ³/₈-inch strap-shaped lip terminating in 3 teeth, the central one much the smallest. Woods, thickets, and meadows.

Spiranthes spp., Ladies' Tresses. These are delicate plants with spikes of small, mostly white, tubular flowers in either a one-sided or spiraling arrangement. Their lips are crisped or toothed at the apex and are not spurred.

> Flowers in 2 or 3 rows: *S. cernua, S. ochroleuca, S. ovalis, S. lucida*
> Flowers in a single row: *S. vernalis, S. lacera, S. tuberosa*

Spiranthes cernua, Nodding Ladies' Tresses (plate 93). This is our most common species and has the largest and most fragrant flowers. There are up to 6 long, narrow basal leaves and a dense 2- or 3-ranked raceme of slightly deflexed white flowers. The lip may be up to ¹/₂ inch long and light yellow green in the center. Wet, grassy open places. Summer–fall. The very similar **S. ochroleuca** (*S. cernua* var. *ochroleuca*) may be distinguished by its pale yellowish flowers.

S. ovalis, Lesser Ladies' Tresses, has a slender raceme with many much smaller white flowers in spiraling rows.

There are a few slender oblanceolate leaves at the base and small bracts on the upper stem. Damp woods. Fall.

S. lucida, Shining Ladies' Tresses, is infrequent in the southern mountains. It has short, blunt oblanceolate leaves and a 3-ranked raceme, but its flowers are slightly larger than in *S. ovalis.* The lip is ¹/₄ inch long with a broad median stripe of bright yellow. Wet places. Late spring–summer.

S. vernalis, Spring Ladies' Tresses. The basal leaves of this species are linear and rigid, persisting through the flowering season. The white flowers have a ¹/₄-inch ovate lip with a yellowish center and are arranged in a single spiral. Open areas. Spring.

S. lacera, Slender Ladies' Tresses, has a basal rosette of ovate leaves usually present at flowering. The inflorescence is loosely secund or spiraled; the flowers are small, with a lip less than ¹/₄ inch long, white but green in the center. In **var. gracilis** (plate 94) the leaves wither before flowering, and the spike is a tight spiral. Open, grassy locations. Summer.

S. tuberosa (*S. beckii, S. grayi*), Little Ladies' Tresses. This is an extremely slender and inconspicuous plant. Its flowers are white (lacking the green center of the preceding species) and tiny (the lip measuring less than ¹/₈ inch) and are in a tight single spiral. The leaves are usually absent at time of flowering. Fields and woods. Summer–fall.

Goodyera pubescens, Downy Rattlesnake Plantain (plate 95). We have two species of *Goodyera*, both notable for their clusters of evergreen, ovate basal leaves which have a network of lighter veins. The flowers are small (under ¹/₄ inch) and are white, but differ from those of *Spiranthes* in having a pouchlike lip prolonged at its apex into a beak. In G. *pubescens* the raceme is densely flowered and cylindrical. The leaves are bluish green, finely reticulated with lighter green, and have a prominent white midvein. Woodlands. Late spring–summer.

G. repens var. ophioides, Lesser Rattlesnake Plantain (plate 96), is a somewhat smaller plant, and its flowers may be tinged with green. The raceme is loosely flowered and secund. Its leaves are dark green with short whitish cross-veins. Cool, damp forests. Summer.

The Rattlesnake Plantains differ from most of our other Orchids in having exceptionally attractive foliage and in being easy to transplant—a combination that has led to their widespread use in terrariums. One official conservation organization reported that more than one-half million individual plants had been collected in Tennessee alone for that purpose.

SAURURACEAE, LIZARD'S TAIL FAMILY

Saururus cernuus, Lizard's Tail, Water Dragon (plate 97), is a plant of swamps and shallow water, spreading from runners to form large colonies. Its leaves are stalked, heart-shaped at the base, and long-pointed. Innumerable minute flowers are crowded into a slender spike up to 6 inches long, which tapers to a gracefully drooping tip. The inflorescence is white owing to the filaments; there are no petals or sepals. Summer.

SANTALACEAE, SANDALWOOD FAMILY

Comandra umbellata, Bastard Toadflax (plate 98). Many of the plants in this family are parasitic even though they contain chlorophyll, and this is no exception. It has numerous elliptic, entire leaves up to 1 $1/4$ inches long, and a terminal cluster of small flowers. The perianth consists of a cuplike calyx with 5 spreading white lobes (there are no petals), each with a stamen inserted at its base. Dry woods and fields. Spring–summer.

ARISTOLOCHIACEAE, BIRTHWORT FAMILY

Aristolochia macrophylla (*A. durior*), Dutchman's Pipe (plate 99). The flowers of our two species of *Aristolochia* are formed by fleshy, tubular S-curved calyxes that terminate in 3 spreading triangular lobes and vary in color from dull greenish yellow to brownish purple. *A. macrophylla* is a climbing, twining vine with heart-shaped leaves up to 8 inches wide. Its flowers are about 1 $1/2$ inches long and usually are borne high above the ground. Rich, rocky woods. Spring.

A. serpentaria, Virginia Snakeroot, is an erect herb up to 2 feet tall and has more elongate leaves with heart-shaped bases. Its flowers, which grow on long stalks near the base of the plant, are similar to those of A. *macrophylla* but are more slender and less than 1 inch long, and have unequal lobes.

Asarum canadense, Wild Ginger (plate 100). In contrast to *Hexastylis* (which is also called Wild Ginger), the leaves of *Asarum* are deciduous and pubescent. In A. *canadense* there are 2 kidney-shaped leaves with cordate bases, about 4 inches wide, on long petioles, and a solitary short-stalked maroon flower arising between the petioles just above ground level. The perianth consists of a bell-shaped calyx about $^3/_8$ inches long, divided at the summit into 3 long-pointed lobes. These lobes are variable, ranging from $^1/_4$ to $^3/_4$ inch in length, and may be spreading or reflexed. Rich woods. Early spring.

Hexastylis arifolia, Little Brown Jugs (plate 101). The foliage of our species of *Hexastylis* is evergreen, glabrous, and often variegated. This species has triangular leaves up to 5 inches long with straight or rounded sides and down-pointing basal lobes. The flowers, which have short, tubular, 3-lobed fleshy calyxes, are on short stalks from the base of the plant. They are pale purplish to olive-brown, about $^3/_4$ inch long, and constricted just below the $^1/_4$-inch spreading lobes. (In **var. ruthii** the lobes are erect and only about $^1/_8$ inch long.) Woodlands. Early spring.

H. virginica, Heartleaf, Wild Ginger. In this and the remaining species, the leaves are ovate or nearly circular with heart-shaped bases. H. *virginica* has a more or less cylindrical calyx tube about $^1/_2$ inch long; the lobes are only $^1/_8$ inch long and spreading. **H. heterophylla** (plate 102) is similar but slightly bell-shaped, and the lobes are $^1/_4$ to $^3/_8$ inch long. Spring.

H. shuttleworthii is our largest species. It has an urn-shaped calyx with a tube between $^3/_4$ and 1 $^1/_4$ inches long and lobes exceeding $^3/_8$ inch. Spring–summer.

H. rhombiformis is a North and South Carolina endemic, found in acidic woodlands. Its calyx is broadly flask-shaped, more or less rhombic, and widest near the middle. Early spring.

POLYGONACEAE, BUCKWHEAT FAMILY

Polygonum spp., Smartweed, Knotweed. The plants in this genus have alternate, entire leaves and small flowers lacking corollas but with 5-lobed calyxes (4-lobed in two species); with few exceptions these are crowded into tight terminal or axillary inflorescences. A distinctive feature, helpful in identifying species, is a cylindrical, membranous sheath (ocrea) that surrounds the stem at each of the swollen leaf joints.

Polygonum cuspidatum, Japanese Knotweed, Mexican Bamboo (plate 103). This and the next two species are large plants of Asiatic origin, often escaping from cultivation. *P. cuspidatum* is a hollow-stemmed perennial that may reach 10 feet in height It has ovate leaves up to 6 inches long, truncate at the base, and erect axillary panicles of whitish flowers. **P. sachalinense** is similar, but its leaves have cordate bases.

P. orientale, Prince's Feather, is a hairy annual up to 8 feet tall with very large, long-stalked, heart-shaped leaves. Its flowers are deep rose, in dense, drooping racemes up to 3 inches long.

P. virginianum *(Tovara virginiana)*, Jumpseed, Virginia Knotweed, differs from most others in having its flowers, which are greenish white and 4-lobed, strung out at intervals along a very slender terminal spike that may be as much as 18 inches long. The leaves are ovate, up to 6 inches long. Woodlands and roadsides. Summer–fall.

P. sagittatum, Arrow-leaved Tearthumb (plate 104). This and the next species have weak, reclining stems armed with tiny retrorse prickles. In *P. sagittatum* the leaves are narrowly arrowhead-shaped, 1 to 3 inches long, with downward-pointing basal lobes. The flowers are white or pink, in small rounded heads. In Halberd-leaved Tearthumb, **P. arifolium,** the leaves are up to 6 inches long, broadly hastate with flaring triangular lobes. Its flowers are pink, in few-flowered racemes. Wet places. Late spring to fall.

P. cilinode, Fringed Bindweed (plate 105). Three of our smooth-stemmed *Polygonums* have trailing or twining habits. *P. cilinode* has ovate leaves with hastate or cordate bases and long-peduncled, branched racemes bearing discontinuous clusters of white flowers. The fringe of bristles from which this plant takes its names are on the bottom edge of the ocrea and point downward. Open areas at high elevations. Summer–fall.

The few flowers of Black Bindweed, **P. convolvulus,** are greenish and are borne in racemes of scattered clusters. The ocreae are without bristles, and the calyxes in fruit are narrowly winged if at all. Disturbed ground. Spring to fall.

P. scandens, Climbing False Buckwheat, has heart-shaped leaves and unbranched racemes. The flowers are white or tinged with pink or green, and the calyx is broadly winged in fruit. There is no fringe on the ocreae. Disturbed areas. Summer–fall.

P. amphibium (*P. coccineum*), Water Smartweed. The remaining species of *Polygonum* are the Smartweeds—erect plants with flowers in axillary and/or terminal spikelike racemes. Water Smartweed usually has a solitary, compact raceme of pink to red flowers, and the ocreae are entire. It may be terrestrial, floating, or submerged, and its adaptations to differing water conditions make it the most variable of our species.

P. pensylvanicum, Pink Smartweed (plate 106), is a tall plant with several erect, cylindrical racemes of light pink flowers and entire or minutely ciliate ocreae. The upper parts of the stem are covered with tiny stalked glands. Disturbed ground. Summer–fall.

P. persicaria, Lady's Thumb, can usually be recognized by the dark blotch near the center of each leaf blade, but sometimes this feature is absent. It is a somewhat sprawling plant with numerous straight, thick ($^3/_8$ inch wide), densely flowered racemes of pink-purple flowers. The summit of the ocrea is fringed, but the cilia are less than $^1/_8$ inch long. Waste ground. Summer–fall.

Long-bristled Smartweed, **P. cespitosum var. longisetum** (plate 107), is similar but without any markings on the leaves, and its racemes are more slender (about $^3/_{16}$ inches wide). The cilia on the ocreae are unmistakable—$^1/_4$ to $^3/_8$ inches long. Its flowers are deep pink. Waste ground. Spring to fall.

P. punctatum is a slender plant in all of its parts. The flowers are white or green and are borne in erect, interrupted racemes in which some of the lower internodes may be 1 inch long. The calyx is conspicuously dotted with glands. Waste areas. Summer–fall.

P. hydropiper, Water Pepper. This also has glandular-punctate calyxes, but its racemes are arched and less interrupted than those of *P. punctatum*. The flowers are greenish with white or pink margins and are 4-lobed. The ocreae have very short cilia. Noted for its acrid, peppery taste. Wet areas. Spring to fall.

Mild Water Pepper, **P. hydropiperoides,** is not markedly peppery. Its racemes are erect and many-flowered (the 5-parted flowers may be pink, green, or white) and the calyxes are eglandular. The ocreae have cilia up to $^3/_8$ inch long and are appressed-hairy.

P. setaceum resembles *P. hydropiperoides*, but has wider (more than $^1/_2$-inch) leaves, densely flowered racemes, and ocreae with longer cilia and spreading hairs. Wet woods and swamps. Summer–fall.

P. lapathifolium, Pale Smartweed, has numerous nodding, slender ($^1/_4$-inch-wide) racemes of pale pink flowers. The ocreae are entire at the summit. Moist, disturbed sites. Spring to fall.

Fagopyrum esculentum (*F. sagittatum*), Buckwheat. More commonly seen as a field crop, Buckwheat also escapes in much smaller numbers to waste areas where it may gain notice as a rather attractive little wildflower. The leaves are triangular with a hastate base, the lower ones stalked (with the characteristic ocreae), the upper sessile. The flowers, which consist of pink or white 5-parted calyxes, are in crowded clusters. Summer–fall.

Eriogonum allenii, Yellow Buckwheat, is our sole member of a genus that is represented by scores of species in the western United States. It is a freely branched, tomentose plant with oblong leaves in whorls. The flat-topped inflorescence contains numerous clusters of flowers each consisting of 6 yellow oval sepals $^1/_8$ inch long, joined at the base. There are no ocreae. Sterile soil, almost exclusively on shale barrens. Summer.

NYCTAGINACEAE, FOUR-O'CLOCK FAMILY

Mirabilis nyctaginea (*Oxybaphus nyctagineus*), Wild Four-o'clock, Umbrellawort (plate 108). This relative of the garden plant sometimes called "Marvel of Peru" may get to be 5 feet high. It has opposite, mostly stalked ovate leaves and small pinkish purple flowers (actually 5-lobed colored calyxes) about $^1/_2$ inch long. Beneath each cluster of a few flowers is a green, lobed, cup-shaped involucre. (Those familiar with tropical plants will be reminded of the related *Bougainvillea*, which has large, brightly colored involucres.) Waste places. Late spring–summer.

PHYTOLACCACEAE, POKEWEED FAMILY

Phytolacca americana, Pokeweed (plate 109), is a rank, widely branched herbaceous plant up to 10 feet tall, with smooth stems strongly tinged with red-purple and large entire leaves. The flowers are borne in 6-inch erect racemes which become deflexed as the black fruit matures. They have 5 round greenish white or pinkish petaloid sepals. Open, disturbed areas. Summer–fall.

Throughout much of Appalachia, Pokeweed is a favorite wild potherb. In spring the very young, tender shoots are cooked in several changes of water; this dish, called "poke sallet," has been favorably if perhaps over-enthusiastically compared to asparagus. Warning: Despite the attractiveness of the plant, all of its other parts, including the mature stems and leaves, are extremely poisonous and must be avoided.

MOLLUGINACEAE, INDIAN CHICKWEED FAMILY

Mollugo verticillata, Carpetweed, is a prostrate, mat-forming weed with whorls of oblanceolate leaves. There are a few $^3/_{16}$-inch apetalous flowers on stalks from each node, the 5 sepals white but green-striped beneath. Summer–fall.

PORTULACACEAE, PURSLANE FAMILY

Claytonia caroliniana, Spring Beauty (plate 110). The appeal of Spring Beauties lies in their pretty flowers, which are $^1/_2$ to $^3/_4$ inch across with pink (sometimes white) petals with darker veins and deep rose anthers. They are short succulent plants with a pair of opposite leaves on the stem below the loose terminal raceme; in

C. caroliniana these are lanceolate to spatulate, usually under 2 ¹/₂ inches in length, with a distinct petiole.

C. virginica (plate 111) is very similar except for the leaves, which tend to be 3 inches long or more and are narrow and long-tapering without an evident petiole. Rich woods. Early spring.

Talinum teretifolium, Fameflower, Rock Portulaca (plate 112), has a cluster of fleshy leaves near the base, very narrow and round in cross-section, and from 1 to 2 inches long. From there arises a wiry stem ending in an open cyme with deep pink flowers that are ¹/₂ inch across. There are 2 sepals, 5 petals, and from 10 to 20 stamens with yellow anthers. These brilliant little flowers open only for a few hours and then only in bright sunshine and usually in mid-afternoon. Rock outcrops and sandy soil. Summer–fall.

Portulaca oleracea, Common Purslane, Pusley. More generally regarded as a potherb or weed than as a wildflower, Purslane does bloom (although only briefly in the morning) from spring until fall. It is a prostrate, branching succulent with reddish stems and thick, flat obovate leaves about 1 inch long. The flowers are mostly solitary in a rosette of leaves, ¹/₄ to ³/₈ inch wide and bright yellow. Fields and waste places as well as gardens.

CARYOPHYLLACEAE, PINK FAMILY

Despite the great diversity in outward appearance among the members of this family, they share a number of common features: The leaves are simple, entire and opposite, and the flowers are radially symmetric with 5 petals in white or shades of red. This description would also seem to fit any of the Phloxes (which are in the family Polemoniaceae), but their petals are joined into a tube whereas in the Caryophyllaceae they are separate.

Sepals united, flowers white: *Silene* (part)
Sepals united, flowers colored: *Silene* (part), *Lychnis,*
  Agrostemma, Saponaria, Dianthus
Sepals separate, styles 5: *Cerastium, Stellaria* (part)
Sepals separate, styles 3: *Stellaria* (part), *Minuartia, Arenaria,*
  Holosteum

Silene noctiflora, Night-flowering Catchfly. Often confused with the next, this can be distinguished by the smaller (³/₄-inch), usually perfect flowers, and by their 3 styles, which is the usual number in this genus. Also, the veins in the calyx are branched. Waste places. Late spring–fall.

S. latifolia ssp. alba (*Lychnis alba*), White or Evening Campion (plate 113), is a hairy plant, sticky-glandular in the upper portion, with an open, branched inflorescence of flowers with bilobed petals, opening in the evening. It is dioecious, the staminate flowers having 10 stamens and the pistillate 5 styles; the downy calyxes (markedly inflated in the pistillate) have 10 and 20 green veins, respectively. Fields and roadsides. Late spring–summer.

S. dichotoma, Forked Catchfly. The stems of this species are hairy but not glandular, and are once or several times dichotomously branched, creating **V**-shaped pairs of long, open, one-sided racemes. The individual flowers (as in the next four *Silenes*) are white, and they resemble those of S. *noctiflora* but are a little smaller and are sessile. The calyxes are narrowly tubular. Road banks and waste places. Spring–summer.

S. vulgaris (*S. cucubalus*), Bladder Campion (plate 114) is a rather weak plant with an open inflorescence. The flowers are ¹/₂ inch wide, with deeply bilobed petals. Very distinctive is the inflated, melon-shaped calyx, which is thin in texture and covered with a network of green or reddish veins. Waste areas. Spring–summer.

S. stellata, Starry Campion (plate 115), has a stiff stem, and, unusual for this genus, most of its leaves are in whorls of 4. The ³/₄-inch flowers have their petals fringed, and the calyx is bowl-shaped. Woods. Summer.

S. ovata, Fringed Campion (plate 116). The petals of this species are not merely fringed but are very deeply cleft into several linear segments. Its calyx is tubular. The plant is coarse and pubescent, the leaves opposite and ovate, with a broad, rounded, somewhat clasping base. Rich woods. Summer–fall.

S. nivea, Snowy Campion. A woodland species less than

1 foot high, Snowy Campion has a few long-stalked flowers
in the upper leaf axils. The petals are only slightly notched
at the apex, and the calyx is slightly inflated and smooth.
Moist woods. Summer.

S. caroliniana ssp. pensylvanica, Wild Pink (plate 117).
This is a low, tufted plant under 10 inches high with
mostly basal leaves. The inflorescence is very sticky and
crowded with pale to deep pink flowers $^3/_4$ to 1 inch
across; the petals are wedge-shaped and either entire or
slightly indented at the apex. Dry, rocky places. Spring–
early summer.

S. virginica, Fire Pink (plate 118), is a stiff, erect plant
with only 2 to 4 pairs of cauline leaves, which are distinctly longer than wide. The flowers are 1 inch across,
the petals bright red, narrow, and toothed or notched.
Open woods and rocky slopes. Spring–summer.

S. armeria, Sweet William, is a smooth annual escaped
from cultivation (but not to be confused with *Dianthus
barbatus*, which is the Sweet William grown as a biennial in gardens). *S. armeria* has sessile or clasping ovate-
lanceolate leaves 1 to 2 inches long, and an inflorescence of congested cymes. The calyxes are $^1/_2$ inch long,
club-shaped, and the petals are about $^1/_4$ inch long,
notched, pink or lavender. Waste places. Summer.

S. antirrhina, Sleepy Catchfly. In terms of flower size,
the *Silenes* reach the lower extreme in Sleepy Catchfly,
where the pink or white corolla measures a mere $^1/_8$ inch
across (or sometimes is lacking altogether) and opens
only briefly in full sun. The plant is quite distinctive,
however—slender, wiry, and divaricately branched, with
widely spaced pairs of narrow leaves and small, reddish
brown glutinous zones in the middle of the upper internodes. The calyxes are ovoid and about $^1/_4$ inch long.
Disturbed ground. Spring–summer.

---

*There is a basic difference in function between the sticky secretions in the
"Catchflies" and other members of the Pink Family, which benefit the plant
only by deterring nonpollinating insects from pilfering nectar, and those of
carnivorous species like the Sundews (Drosera), which play an active role in
obtaining supplementary nutrients.*

Lychnis flos-cuculi, Ragged Robin (plate 119), is an occasional escape from cultivation. Its purplish pink petals are deeply divided into 2 narrow lobes and have a pair of short, slender lateral appendages. The calyx tube is not inflated. Summer.

Agrostemma githago, Corn Cockle (plate 120). This handsome plant is actually a noxious European weed that has spread to fields and waste places here. It has a few pairs of linear leaves, and solitary flowers terminating the branches. They are 1 $^1/_2$ inches across, with red-purple petals; there are 5 styles. The calyx is 10-ribbed and has 5 slender lobes that protrude far beyond the corolla. Spring–fall.

Saponaria officinalis, Bouncing Bet, Soapwort (plate 121). A smooth, stout plant from 1 to 2 feet high, Bouncing Bet is a persistent weed in waste places and disturbed ground, especially on railroad embankments. It gets its other names from its lather-forming sap. The inflorescence is crowded with pink to nearly white flowers, the petals narrowly wedge-shaped, indented at the summit, and somewhat reflexed. There are 2 styles. Spring–fall.

This immigrant from the Old World has an unusually high content of saponin, a substance which combines with water to form frothy lather. The usefulness of this plant in the washing of clothes was well known to the colonists, who brought it to America and placed it under cultivation for this as well as medicinal purposes, and as a garden flower.

Dianthus armeria, Deptford Pink (plate 122). The flowers of Deptford Pink are less than $^1/_2$ inch wide, but they compensate for lack of size with their vividly colored petals, which are rose-pink copiously dotted with white and have a few teeth on the margins. They are in congested cymes together with long, erect awl-shaped bracts. The leaves are numerous, very narrow, and ascending. Dry fields and waste places. Spring–summer.

Cerastium arvense, Field Chickweed (plate 123). All of the remaining plants in the Pink Family have white flowers. In *Cerastium* the petals are seldom entire but are less deeply cut than in *Stellaria*. This species is an attractive weed of abandoned fields and rocky places but is infrequent in the southern mountains. It has linear to narrowly lanceolate leaves, with leafy shoots present in some axils. The flowers are $^1/_2$ inch wide and the petals obcordate, much longer than the sepals. Spring–summer.

C. fontanum ssp. triviale (*C. vulgatum, C. holosteoides* var. *vulgare*), Mouse-ear Chickweed (plate 124). This is a common weed of lawns and waste areas. The stem is hairy and sticky and the sessile leaves oblong, usually under 1 inch. There are numerous flowers, $^1/_4$ inch wide, the petals cleft and about equaling the sepals. Spring–fall.

C. glomeratum (*C. viscosum*) is another common and widespread weed in fields, lawns, and waste places. A tufted plant with many viscid-hairy stems up to 1 foot tall, it has clasping spatulate leaves 1 inch long and about half as wide. The inflorescences are compact and many-flowered; the petals and sepals are each $^3/_{16}$ inch long. Spring–early summer.

C. semidecandrum is similar but smaller in all respects and with less-congested inflorescences. There often are 5 stamens instead of the usual 10. Disturbed sites. Spring–early summer.

C. brachypetalum has an open, dichotomously branched inflorescence with numerous flowers. The petals and sepals are about equal, only $^1/_8$ inch long. Disturbed sites. Spring–early summer.

C. nutans is our only native *Cerastium* and tends more toward woodland habitats than do the others. It is weak and sprawling, with narrow leaves up to 2 inches long. The inflorescences are open, and the petals $^1/_4$ to $^3/_8$ inch long, much exceeding the short sepals.

Stellaria aquatica (*Myosoton aquaticum*), Water Chickweed. Its 5 styles set this species apart from the other members of its genus, which have 3. It is a prostrate plant of wet areas, with sessile ovate leaves, and flowers in open, leafy terminal cymes. The petals are about $^1/_4$ inch long and cleft almost to the base. Spring to fall.

S. pubera, Giant or Star Chickweed (plate 125). This attractive woodland species is certainly regarded as being more a wildflower and less a weed than the others.

Its flowers are $^1/_2$ inch wide, and, as with most *Stellarias*, the petals are cleft almost to the base, making them appear to be 10; in this species they exceed the sepals. It is an upright plant with sessile elliptic leaves up to 2 $^1/_2$ inches long. The very similar **S. corei,** known as Tennessee Chickweed, has acuminate sepals that are longer than the petals, and some of its leaves may be petiolate. Spring–early summer.

S. media, Common Chickweed. This ubiquitous plant is one of our best-known weeds, thriving in gardens and other disturbed locations and flowering at nearly all times of the year. It is weak and prostrate, with ovate leaves on distinct stalks and flowers less than $^1/_4$ inch wide, the petals exceeded by the sepals. A peculiarity is that they may have fewer than the usual complement of 10 stamens.

S. graminea, Lesser Stitchwort (plate 126). The Stitchworts are smooth, wiry-stemmed plants with narrow leaves, usually found in grassy places. *S. graminea* has an open, many-flowered, divaricately branched terminal inflorescence. The flowers are about $^3/_8$ inch wide, with petals longer than the sepals. Long-leaved Stitchwort, **S. longifolia,** on the other hand, has only a few flowers in lateral, ascending inflorescences. Spring–fall.

S. alsine, Bog Chickweed. A plant of wet habitats, *S. alsine* is decumbent, rooting at the nodes, with lanceolate leaves. Its flowers are $^1/_4$ inch across, in short cymes. Summer–fall.

Minuartia groenlandica (*Arenaria groenlandica*), Mountain Sandwort (plate 127). In contrast to *Stellaria*, the petals in this genus are either entire or shallowly indented at the apex. M. *groenlandica* is a smooth perennial with tufts of basal leaves and slender, erect stems bearing opposite linear, blunt-pointed $^1/_2$- to $^3/_4$-inch leaves and cymes of $^1/_2$-inch flowers. The petals are notched and are longer than the sepals.

M. glabra is similar but is an annual without basal leaves at time of flowering. Rock outcrops. Spring–summer.

M. uniflora (*Arenaria uniflora*) is a delicate, wiry plant that rises only a few inches from shallow soil on granite outcrops, where its habit of forming dense mats renders it conspicuous despite its diminutive stature. It has a rosette of basal leaves, and stem leaves under ¹/₄ inch long. The flowers are ³/₈ inch across. Very rare and local. Spring.

M. stricta (*Arenaria stricta*), Rock Sandwort, also has narrow leaves, but unlike those of the preceding species they are rigid, and have fascicles of secondary leaves in their axils. The flowers are about ³/₈ inch across with entire petals, in an open forked inflorescence. Rocky soil. Summer.

Arenaria serpyllifolia, Thyme-leaved Sandwort. In this much-branched species, the leaves are ovate and less than ¹/₄ inch long. The flowers are also small—only ¹/₈ inch wide and exceeded by the needle-pointed sepals. Fields and waste places. Spring–summer.

Holosteum umbellatum, Jagged Chickweed. This plant is unusual in that its flowers are in umbels at the summit of an unbranched stem; the flowers are very small. Most of its leaves are basal. Disturbed areas. Spring.

NYMPHAEACEAE, WATER LILY FAMILY

Nymphaea odorata, Fragrant Water Lily (plate 128). This is probably the most familiar of our aquatic plants. Its floating leaves, which can measure up to 10 inches, are almost circular with a deep V-shaped sinus and are purple beneath. The solitary 3- to 5-inch flower has numerous sharp-pointed white (rarely pinkish) petals that transform gradually and almost imperceptibly into stamens near the center. It opens in the mornings for a few successive days. Quiet water. Summer.

Nuphar luteum ssp. macrophyllum, Spatterdock, Yellow Pond Lily (plate 129). These plants have large, oval leaves notched at the base, either floating or held above the surface. The conspicuous parts of the 1- to 2-inch, bowl-shaped flower are the 6 convex yellow sepals, which almost conceal the many-lobed stigma. Spring–fall.

Brasenia schreberi, Water Shield, is an aquatic plant with branching stems that bear crowded masses of floating leaves. These are oval, peltate, and up to 4 inches long. The flowers, which are held erect above the surface on thick stalks, have 3 sepals and 3 petals, dull purple and about ³/₄ inch long, and with 12 to 20 prominent stamens. In ponds and slow-moving water. Early summer–fall.

RANUNCULACEAE, CROWFOOT FAMILY

The first few genera in this family have bilaterally symmetric (zygomorphic) flowers. The remainder can be divided between 1) those in which the most conspicuous part of the flower consists of petals or showy sepals, and 2) those in which the perianth parts are either inconspicuous or absent, leaving it to the numerous stamens and/or pistils to constitute what we tend to perceive as a flower.

> Petals or sepals conspicuous; flowers zygomorphic: *Aconitum*, *Delphinium*, *Consolida*
> Petals or sepals conspicuous; flowers actinomorphic: *Xanthorhiza*, *Aquilegia*, *Hepatica*, *Clematis*, *Anemone*, *Enemion*, *Coptis*, *Caltha*, *Ranunculus*, *Thalictrum* (part)
> Petals or sepals inconspicuous or absent: *Hydrastis*, *Actaea*, *Cimicifuga*, *Trautvetteria*, *Thalictrum* (part)

Perianth conspicuous; flowers zygomorphic

Aconitum uncinatum, Monkshood (plate 130), has a weak, often climbing stem with numerous leaves palmately lobed into 3 to 5 coarsely toothed segments. The few dark blue flowers are formed by their irregular calyxes; the upper sepal (called the helmet in this genus) is enlarged into a hood about ³/₄ inch high, concealing 2 small petals (the other 3 are minute or may be absent altogether). Rich woods. Summer–fall.

A. reclinatum, Wolfsbane. Despite the specific epithet, this is usually erect. The flowers are more numerous than in the preceding and are white or pale yellow. The hood is horizontal and elongated. Rich woods. Summer–fall.

Delphinium tricorne, Dwarf Larkspur (plate 131). The Larkspurs differ from *Aconitum* in that all 5 sepals have petal-like blades somewhat similar in color and shape except that the upper one is prolonged backward into a long, hollow spur. *D. tricorne* is a spring-blooming species usually less than 18 inches high. It has a loose raceme of dark blue to violet or sometimes white flowers; the sepals are about ¹/₂ inch long, the spur much longer. Its leaves are deeply divided into several narrow-lobed segments.

D. exaltatum, Tall Larkspur, may attain a height of 6 feet and has numerous blue or white flowers in elongate racemes, with sepals about ³/₈ inch long. The leaves are cleft into a few lanceolate divisions. Unlike the preceding species, it flowers in summer or early fall.

Consolida ambigua (*Delphinium ajacis, D. ambiguum*), Larkspur. This European annual Larkspur may be found along roadsides and in waste places during the summer. It can be readily identified by its leaves, which are dissected into very narrow linear segments.

*Perianth conspicuous; flowers actinomorphic
(See also Thalictrum)*

Xanthorhiza simplicissima, Yellowroot (plate 132). This is a low shrub with bright yellow inner bark, often forming a dense growth along stream banks. Several pinnately divided, toothed leaves emerge in early spring at the top of the stem along with drooping racemes of flowers. These are star-shaped, no more than ¹/₄ inch wide, of 5 brownish purple petaloid sepals.

Yellowroot has been used in a number of backwoods remedies, including a rather startling prescription for a concoction of the bitter root in whiskey together with sourwood honey, to be taken for stomach ulcers. Today the basketmakers of the Eastern Band of Cherokees crush the entire plant to obtain a bright yellow dye for the split white oak baskets for which they are famous.

Aquilegia canadensis, Columbine (plate 133). This is our only representative of the Crowfoot Family with flowers that are both radially symmetric and spurred. The only Columbine native to the East, it cannot be mistaken for any other wildflower here. The 1 ¹/₂-inch, red-and-yellow flowers dangle from wiry stems, with a column of stamens protruding downward and a long, curved spur extending backward from each of the 5 petals. Foliage is divided and redivided into threes. Rocky woods and slopes. Spring–summer.

Hepatica americana (*H. nobilis* var. *obtusa*), Round-lobed Hepatica, Liverleaf (plate 134). Hepaticas are among the delightful early spring wildflowers of our deciduous woods. They are the only plants in the Ranunculaceae in which all of the leaves are basal; the blossoms appear among the previous year's leaves and before the new ones are fully developed. What appear to be petals are actually sepals, and beneath these are 3 bracts simulating a calyx. In *H. americana,* both these bracts and the 3 leaf lobes are rounded or blunt. The flowers are between ¹/₂ and 1 inch across, and may be blue, violet, pink, or sometimes white.

As might be expected from the name, Sharp-lobed Hepatica, **H. acutiloba** (*H. nobilis* var. *acuta*) (plate 135), has acute leaf lobes and bracts. White flowers are common and deep colors somewhat less frequent.

The name "Hepatica" is from the Greek word for liver, and apparently was chosen because of a supposed resemblance of the leaf shape, as well as the dark red-brown color of the old foliage, to that organ. It is not surprising, therefore, that old-time herbalists recommended that these plants be used in treating liver disorders.

Clematis virginiana, Virgin's Bower (plate 136). This genus is unique within the Ranunculaceae in having opposite leaves. Its flowers lack petals. *C. virginiana* and the next following species, both climbing vines, are the only ones in our area with numerous flowers in axillary clusters. Each has 4 white, oval, spreading sepals about ¹/₂ inch long. The leaves are pinnately compound, those in this species having 3 coarsely toothed segments on stalks of equal length. Thickets and roadsides. Summer.

C. terniflora (*C. dioscoreifolia*), an alien introduction often escaping from cultivation, is similar but with 3 or 5 entire leaflets, the terminal one long-stalked.

C. occidentalis (*C. verticillaris*), Mountain Clematis, is a vine with large (up to 2-inch) dull blue flowers solitary in the leaf axils. They are markedly different from the succeeding species in that the 4 sepals are thin in texture and are separate. Rocky woods. Spring–early summer.

C. viorna, Leatherflower (plate 137). The remaining species have very thick, fleshy sepals joined at the base to form an urn-shaped flower. *C. viorna* is a vine with ³/₄- to 1-inch flowers on long stalks in the axils. Its sepals are white with a purplish red outer surface and recurved tips. Rich woods. Late spring–summer.

C. ochroleuca, Curlyheads. This is the only erect herb in this genus likely to be encountered in our region. Its leaves are ovate and sessile, and the flowers solitary. The sepals are about $^3/_4$ inch long, yellowish white, and recurved at the apex. The plumose styles are 1 to 2 inches long and tawny at maturity. Dry, rocky soil. Spring–early summer.

Two other species, both shale-barren endemics, are to be found in our area, but only in a very few localities in Virginia. **C. albicoma** has whitish or yellowish styles; in **C. viticaulis** they are short and dark brown.

Anemone spp. The *Anemones* are characterized by the combination of white flowers (of 5 or more sepals) and a whorl of involucral leaves (usually 3) subtending the inflorescence. The first two species are well over 1 foot tall and may bear one to three flowers.

Anemone virginiana, Thimbleweed (plate 138), has leaves that are deeply parted into 3 sharply incised segments, and all, both basal and involucral, are stalked. The flowers are greenish white, $^3/_4$ inch or more wide. Dry woods, rocky banks. Spring–summer.

In **A. canadensis,** Canada Anemone (plate 139), the leaves are less deeply cut and those of the involucre are sessile. The flowers are white and up to 1 $^1/_2$ inches across. Moist meadows and low woods. Spring–summer.

Two less common species are about 1 foot high and have solitary flowers with 10 or more narrow sepals, which may be purplish tinged. **A. caroliniana** has leaves that are divided into 3 wedge-shaped, lobed segments, and the involucre is below the middle of the scape. In **A. berlandieri** the involucral leaves are well above the middle and are divided into very narrow, almost linear lobes. Both bloom in spring.

A. quinquefolia, Wood Anemone (plate 140), is a small, delicate plant not over 8 inches high, with a solitary flower of 5 or 6 sepals about $^1/_2$ inch long, purplish on the back. The involucral leaves are long-stalked and deeply divided into 5 segments, or into 3 with the lateral ones deeply incised. **A. lancifolia,** Mountain Anemone, is similar but larger (up to 1 foot high), the leaf segments only 3 and the lateral ones not cut, and the sepals $^3/_4$ inch long. Rich woods. Spring.

Enemion biternatum *(Isopyrum biternatum)*, False Rue Anemone, barely enters our area. Its flowers are white and consistently 5-merous. The leaflets are more deeply lobed than those of *Thalictrum thalictroides*. The leaves are basal and cauline, biternate below but once-ternate above (there is no involucre); the flowers are axillary and terminal. Rich woods. Early spring.

Coptis groenlandica *(C. trifolia)* (plate 141) is known as Goldthread because of its bright yellow slender rhizome, which sends up long-stalked leaves and a solitary white scapose flower. The foliage is evergreen and shiny, each leaf having 3 roundish toothed leaflets. The $^1/_2$-inch flowers have 5 to 7 narrow petaloid sepals, numerous stamens, and 3 or more stalked pistils topped by a persistent style. Damp woods and bogs. Spring.

Caltha palustris, Marsh Marigold (plate 142). All of the remaining species described in this section have alternate leaves and shiny yellow flowers. In *Caltha,* which is represented by a single species, the colorful parts are sepals; in the rest, which belong to the genus *Ranunculus*, it is the petals that are yellow. In our area Marsh Marigold is found only in the northern portion, in shallow water or wet ground. Its leaves are almost round, cordate at the base, toothed but never divided or cleft, from 2 to 6 inches long. There are many bright yellow flowers up to 1 $^1/_2$ inches across. Spring–early summer.

For some reason we in America have given Caltha palustris *a second common name, "Cowslip," despite the fact that this already belonged to Primroses* (Primula) *in England where both are native. "Marsh Marigold" is accepted in both hemispheres—but then it is not a Marigold either.*

Ranunculus ambigens, Spearwort. Spearworts are weak-stemmed, small-flowered plants with simple leaves, growing in muddy soil. This species roots at the nodes and has lanceolate, slightly toothed alternate leaves. The petals are $^1/_4$ inch long, slightly exceeding the sepals. **R. laxicaulis** and **R. pusillus** are erect and have long-stalked ovate or lanceolate basal leaves and linear cauline leaves; the first has 5 petals $^1/_8$ to $^3/_8$ inch long (much surpassing the sepals), the second 1 to 3 petals only $^1/_{16}$ inch long. Spring–summer.

R. abortivus, Kidney-leaved Crowfoot (plate 143). Several species of Crowfoot have simple basal leaves and divided cauline leaves. In R. *abortivus* some of the basal leaves are kidney-shaped with scalloped margins, and the cauline leaves are divided into 3 narrow, blunt-lobed segments. The stem and leaves are smooth. Its petals are less than $^1/_8$ inch long, smaller than the sepals. Mountain Crowfoot, **R. allegheniensis,** is very similar; it can best be distinguished by the beak of the achene, which is longer and strongly curved. **R. micranthus** differs from both in being villous rather than smooth. A fourth species, **R. harveyi,** has $^1/_4$-inch obovate petals—twice as long as the sepals. Spring–early summer.

R. sceleratus, Cursed Crowfoot. In this and the succeeding species, all of the leaves are deeply lobed or divided. Cursed Crowfoot is a smooth, succulent plant of wet habitats with $^1/_4$-inch flowers.

R. parviflorus is a weed in lawns and waste areas, soft-hairy, and with flowers that are only $^1/_8$ inch across. Spring–summer.

R. recurvatus, Hooked Buttercup (plate 144). The flowers of this species are only $^3/_8$ inch wide. The common name derives from the beak of the achene, which is strongly hooked. The leaves are hairy, the principal ones nearly round in general outline and deeply 3-lobed into broad segments, and are up to 2 $^1/_2$ inches wide. Woodlands. Spring–summer.

R. bulbosus, Bulbous Buttercup. This species may be recognized by its hairy stem, which is erect and bulbously thickened at its base. The flowers are about $^3/_4$ inch wide, and the sepals are sharply reflexed. Its leaves are deeply cleft and lobed, the segments broad and blunt-toothed. Open areas. Spring–summer.

R. repens, Creeping Buttercup. This is a hairy, creeping plant, rooting at the nodes and often forming extensive mats. The flowers are $^3/_4$ inch across, sometimes having numerous petals. The leaves are deeply cleft into sharply toothed segments and are often mottled with lighter green. Moist, open places. Spring–early summer.

R. acris, Tall Buttercup (plate 145). This is a very conspicuous plant, attaining a height of 3 feet and often covering fields with blankets of brilliant yellow. It is slender, with leaves up to 4 inches wide, palmately divided into 5 or 7 unstalked narrow segments. The flowers are $^3/_4$ inch wide. Spring–fall.

R. hispidus, Hispid Buttercup (plate 146). The remaining species constitute a perplexing polymorphic group. Ideally, identification should be attempted only when fully developed achenes are present, and with the help of a technical key. The following descriptions merely recite some of the characters that have been ascribed, with less than total unanimity, to the various species. Hispid Buttercup is the most common in our region. It is erect and hairy, with leaves broader than long but extremely variable. The petals are about $^3/_8$ inch long.

Early Buttercup, **R. fascicularis,** has a more limited distribution in the southern mountains. It is similar, but its leaves are longer than wide and the segments narrower. The petals also are longer than broad. Both of these species are usually found in dry habitats. Early–late spring.

R. septentrionalis, known as Swamp Buttercup (plate 147), and **R. carolinianus** are virtually smooth and have reclining stems. The first is stout, the largest leaves 4 inches wide, and has $^1/_2$-inch obovate petals and spreading sepals, while the second is slender, the petals smaller and narrow, and the sepals reflexed. Both inhabit moist woods and swamps, but neither is plentiful in our area. Spring.

Perianth inconspicuous or absent

Hydrastis canadensis, Golden Seal, Orangeroot (plate 148), is unusual for this group in having a solitary flower. The plant is 1 foot high with a large basal leaf and a stem bearing 2 smaller cauline leaves, all deeply palmately lobed and toothed. The single terminal flower consists of numerous white stamens surrounding the pistils, about $^1/_2$ inch across. Rich woods. Spring.

Considered one of the most versatile medicinal plants in the Appalachians, Golden Seal has been collected to the point where its survival may be in jeopardy. Long before this became a concern, indigenous peoples were mixing its powdered rhizomes with bear grease to make an insect repellent.

Actaea pachypoda (A. *alba*), White Baneberry, Doll's Eyes (plate 149 and plate 150). White Baneberry is usually 2 feet high and has 2 large compound leaves with sharply toothed leaflets and a long-stalked 1- to 2-inch oblong simple raceme of white flowers made conspicuous by their numerous long stamens (the perianth segments are very narrow and fall early). The fruits are white berries with a purplish black "eye," on thick, fleshy, bright pink pedicels. Rich woods. Spring.

Cimicifuga racemosa, Black Cohosh, Black Snakeroot, Bugbane (plate 151). In contrast with *Actaea*, the plants in this genus have one or more slender, elongate racemes of white flowers which may exceed 1 foot in length. The leaves are large, decompound, and coarsely toothed. In this species each flower has a single sessile pistil with a thick style and a blunt stigma. Rich woods. Summer.

Mountain Bugbane, **C. americana** is similar to the preceding but the flowers have 3 or more stalked pistils with a slender, curved style. Less widespread, usually at high elevations.

Trautvetteria carolinensis, Tassel Rue, False Bugbane (plate 152), is a tall plant with its principal leaves up to 8 inches wide, palmately cleft into 5 to 11 wedge-shaped segments with large sharp teeth. The basal leaves are long-stalked, the cauline ones much smaller and sessile. The flowers are $1/2$ inch wide, with numerous white stamens, in a corymbiform cluster. Along stream banks and in other moist areas. Summer.

Thalictrum spp., Meadow Rue. All of our plants in this genus have ternately compound leaves with leaflets that are small and shallowly lobed across their summits and (with one exception) branched panicles of flowers in which the sepals are small and not petal-like.

Thalictrum clavatum, Mountain Meadow Rue, Lady Rue (plate 153), is the only species with perfect flowers—i.e., each containing both pistils and stamens. The filaments of the stamens are white and greatly dilated, suggesting club-shaped petals. Wet, shady places. Spring–summer.

T. dioicum, Early Meadow Rue (plate 154). In this and the remaining species, the stamens and pistils are in separate flowers, and these usually are on separate plants. Identification is most easily accomplished by examining the staminate flowers. *T. dioicum* comes into flower in early spring, before any of the others. Its stamens are $^1/_4$ inch long, threadlike, and yellowish, in pendulous clusters. All of its leaves have long petioles, and the leaflets may have as many as 9 rounded lobes. Rich woods.

T. coriaceum also has filiform, pendent yellowish stamens, $^3/_8$ inch long. It is an erect plant with short-stalked leaves; the leaflets have 4 or more lobes. **T. steeleanum** is similar but is stoloniferous, and the terminal leaflets are wider than long. Rich woods and thickets. Late spring–summer.

T. revolutum, Waxy or Skunk Meadow Rue. This species also has drooping stamens, but they have white filaments slightly thickened toward the anthers. The foliage is malodorous, the leaflets entire to 3-lobed, with revolute margins and minute, waxy glands beneath. Woodlands. Late spring–summer.

T. pubescens (*T. polygamum*), Tall Meadow Rue (plate 155). In Tall Meadow Rue the filaments are white and distinctly club-shaped, but are erect and radiate in all directions to form globose clusters. Its leaflets are usually 3-lobed and more than $^1/_2$ inch long. **T. macrostylum** is superficially similar but has smaller leaflets, most of which are entire. Moist habitats. Summer.

T. thalictroides (*Anemonella thalictroides*), Rue Anemone (plate 156), has flowers that bear a much closer resemblance to those of the *Anemones* than to any of our other species in *Thalictrum*. They are borne 3 or more in an umbel, with 5 to 10 sepals about $^1/_2$ inch long, white or occasionally pink. The segments of its biternately compound leaves are small, roundish, and shallowly lobed; it also has an involucre of 2 sessile ternately compound leaves. Rich woods. Early spring.

BERBERIDACEAE, BARBERRY FAMILY

Podophyllum peltatum, May Apple, Mandrake (plate 157). Flowering plants of May Apple have a pair of leaves 6 to 8 inches wide, deeply and radially 5- to 9-lobed. A solitary waxy white flower nods on a short stalk from the junction of the two leafstalks. An inch or two wide, it has 6 to 9 petals and twice as many stamens. Woods and moist clearings. Spring.

Although American Indians used May Apple for a variety of ailments (and were emulated in this by colonial herb doctors) and despite the fact that the modern medicine podophyllin is obtained from its root, the entire plant is considered poisonous. Least toxic is the fully ripe fruit, which many find safe to eat. It is sometimes called "Mandrake," but this name properly belongs to another toxic but unrelated plant native to the Mediterranean area.

Diphylleia cymosa, Umbrella Leaf (plate 158), also has 2 stalked leaves, but they are alternate, the upper slightly smaller but each approaching 2 feet in width. They are 2-cleft and coarsely dentate. The numerous flowers are $^1/_2$ to $^3/_4$ inch wide, with 6 white petals, and are held above the foliage in a long-stalked cyme. The fruits that follow are blue berries on bright red stems. Seeps in cool mountain forests. Spring.

Jeffersonia diphylla, Twinleaf (plate 159). The few basal leaves of *Jeffersonia* are deeply divided into 2 segments. A solitary flower is borne on a naked stalk; with 8 white petals about $^3/_4$ inch long, it bears a resemblance to that of the much more common Bloodroot. Woods, especially in basic soils. Early spring.

Caulophyllum thalictroides, Blue Cohosh (plate 160), is a smooth, glaucous herb up to 3 feet high, bearing one very large compound leaf that is divided and redivided so that it looks very much like many small leaves. There are a number of $^1/_2$-inch flowers in a loose panicle. What appear to be 6 greenish yellow or brownish purple petals are actually sepals; the true petals appear at the inner bases of these as small yellowish glands. To make the deception complete, the bright blue "fruits" that give it its popular name are really berrylike seeds. Rich woods. Spring.

PAPAVERACEAE, POPPY FAMILY

Papaver dubium, Poppy (plate 161). This Poppy has escaped from cultivation to fields and roadsides, where it blooms from spring into summer. It has alternate leaves that are pinnately divided into coarsely toothed segments; it is beset with stiff hairs and contains milky sap. The flower has 4 thin-textured, scarlet petals 1 inch long. The stigma is a radially lobed disk (5- to 9-rayed in this species) on the summit of the ovary.

Chelidonium majus, Celandine (plate 162). Another escape, this one prefers moist woodland margins. It is smooth, with alternate leaves pinnately and irregularly lobed, and small umbels bearing $3/4$-inch yellow 4-petaled flowers. Its juice is deep yellow. Spring–fall.

Stylophorum diphyllum, Celandine Poppy (plate 163). Their common names are frequently the cause of confusion between this and *Chelidonium,* and in fact there are a number of similarities in their leaves, flowers, and juices. *Stylophorum* is somewhat larger, however, with a single or very few flowers about 2 inches wide. There are basal leaves, but more notable is a pair of smaller ones high on the stem. Damp woods. Spring.

Sanguinaria canadensis, Bloodroot (plate 164). The red-orange juice that gives Bloodroot its name also hints at its membership in the Poppy Family. The solitary white flower, which is 1 to 1 $1/2$ inch across, normally has 8 petals (4 slightly longer than and alternating with the others), but sometimes there are more. Its single leaf is circular in outline but palmately lobed, the underside glaucous and conspicuously veined; it is folded around the flower scape and does not fully expand until after blooming. Rich woods. Early spring.

As early as the settlement of Jamestown, native tribes were known to have used the colorful juice of Bloodroot to decorate their faces and bodies for religious ceremonies, and today's children are tempted to imitate them, but this is a dangerous practice due to the presence of poisonous alkaloids. The plant has long been used medicinally, however, and is the subject of much current research.

Argemone mexicana, Prickly Poppy (plate 165), is a native of tropical America occasionally escaping from cultivation. It has spiny foliage, 4- to 6-petaled yellow flowers up to 2 ¹/₂ inches wide, and yellow latex. A western species, **A. albiflora** *(A. alba),* is similar but has larger, white flowers and colorless latex. Both are found in dry waste areas and may bloom any time between spring and fall.

FUMARIACEAE, FUMITORY FAMILY

Dicentra cucullaria, Dutchman's Breeches (plate 166). Our plants in this genus have an unusual flower structure known as bisbilateral, which means that the corolla is divisible into equal halves in two planes rather than only one as, for example, in *Corydalis*. There are 4 petals, the outer 2 dilated at their bases into a pair of spurs, the inner ones much less conspicuous. *D. cucullaria* is usually less than 1 foot high and has finely dissected, fernlike foliage. The white, sometimes pinkish, flowers are about ³/₄ inch long and dangle in a raceme. The flattened corolla is divided at the base into 2 tapering, divergent spurs pointing upward (the flowers hang upside down) and is yellow-tinged at the summit.

Squirrel Corn, **D. canadensis** (plate 167), is similar except that the spurs are rounded, creating a heart-shaped base. Both are woodland species blooming in early spring.

D. eximia, Wild Bleeding Heart (plate 168). This will be recognized by many as a close relative of the cultivated Asian *D. spectabilis*. It is no more than 2 feet tall, with leaves more coarsely divided than in the preceding two species. Its flowers, which are borne in a short panicle, are deep pink, about 1 inch long, and are shaped like those of *D. canadensis* but with longer tips on the outer petals. Rocky woods and ledges. Spring.

Adlumia fungosa, Climbing Fumitory, Allegheny Vine (plate 169). This is a delicate clambering vine with twining leafstalks and ternately compound leaves. Its flowers are pale pink-purple, ³/₄ inch long and only one-third as wide, with a rounded base. *Adlumia* blooms in summer and fall, but the persistent corollas can often be found much later, spongy in texture but little changed in outward appearance. Rocky slopes.

Corydalis flavula, Yellow Fumitory (plate 170). We have two yellow-flowered species of *Corydalis*, both about 1 foot high with bipinnately dissected leaves and small racemes of bilaterally symmetric flowers with a blunt, dilated spur at the base of the corolla. *C. flavula* has ³/₈-inch flowers with a short spur that is turned downward at the end; the crest of the upper petal is toothed. The racemes are about equal to the leaves. Rocky woods. Spring.

In Slender Corydalis, **C. *micrantha* ssp. *australis,*** which is scarcer in our mountains, the racemes extend well beyond the leaves, the flowers are larger (¹/₂ inch), the spur is longer and either straight or turned upward, and the upper petal is untoothed. Rocky slopes. Spring.

A third species, Pale Corydalis, **C. *sempervirens*** (plate 171), grows to 2 feet or more in height and is found on open cliffs and rock outcrops. It is a glaucous plant and bears pink flowers with yellow tips, ¹/₂ to ³/₄ inch long. Spring–summer.

Brassicaceae, Mustard Family

Plants in the Mustard Family can easily be distinguished from others with 4-petaled flowers by their stamens, of which there are six, 2 of them shorter than the other 4. Identifying the species is much more difficult and may require examination of the fruits, but fortunately some seed pods usually are present along with the flowers. More than half of our species are firmly established as weeds and are common in fields and waste areas.

> Flowers yellow; some leaves lobed: *Barbarea, Brassica, Raphanus, Sisymbrium, Rorippa*
> Flowers yellow; no leaves lobed: *Isatis, Erysimum*

Flowers white; pods about as wide as long; petals under $^1/_8$
  inch: *Capsella, Thlaspi, Lepidium*
Flowers white; pods longer than wide; petals under $^3/_8$ inch:
  Arabidopsis, Draba, Arabis, Sibara, Cardamine (part),
  Nasturtium, Alliaria
Flowers white, pink, or purple; pods longer than wide; petals
  over $^3/_8$ inch: *Cardamine* (part), *Hesperis, Lunaria*

Flowers yellow

Barbarea verna, Early Winter Cress. Our two species
of *Barbarea* come into bloom in early spring in fields
and waste areas. Their yellow flowers are about $^1/_4$ inch
wide. Both have pinnatifid basal leaves with a large terminal lobe. In this species these also have from 4 to 10
pairs of smaller lateral lobes; the cauline leaves are similar but much reduced. Its seed pods are 1 $^1/_2$ inches long
or more.

B. vulgaris (*B. arcuata*), Winter Cress, Yellow Rocket
(plate 172), is similar but its basal leaves have only 1 to
4 pairs of lateral lobes. Its pods seldom exceed 1 inch in
length.

Without question, the favorite native potherb in the southern mountains is Barbarea
vulgaris (*with* B. verna *a close runner-up*). *It has even made it to the canned
vegetable shelves of grocery stores under the name of "Creasy Greens."*

Brassica napus (*B. campestris, B. rapa*), Field Mustard.
This genus can be distinguished from the preceding one
by the long ($^1/_4$- to $^1/_2$-inch) beaks on their pods, compared with less than $^1/_8$ inch in *Barbarea*, and by their
flowers, which are about $^1/_2$ inch wide. In *B. napus* some
of the leaves are entire or merely toothed, and auriculate-clasping.

B. juncea, Indian Mustard, is less common in the mountains. Its stem leaves may be sessile or short-petiolate
but never are clasping. Fields and waste ground. Spring–
summer.

Raphanus raphanistrum, Wild Radish (plate 173), has
distinctive flowers, $^1/_2$ to $^3/_4$ inch wide with yellowish
petals veined with lilac but turning white with age. The
long-beaked pods are up to 1 $^1/_2$ inches long, and when

mature become constricted between the seeds. A weed in waste places. Spring–summer.

Sisymbrium officinale* var. *leiocarpum, Hedge Mustard, has some of its cauline leaves deeply cut into 3 or 5 sharp-pointed, coarsely toothed segments, the basal lobes spreading widely. The flowers are only $^1/_8$ inch wide, and the pods $^1/_2$ inch long and closely appressed to the stem. Fields and waste areas. Summer.

In **S. *altissimum,*** Tumble Mustard, the cauline leaves are pinnately parted into 5 to 8 pairs of toothed segments, the upper ones linear and entire. Its flowers are slightly larger, and the pods are much longer and not appressed. Waste areas. Spring–summer.

Rorippa palustris *(R. islandica),* Yellow Cress. This is a small species, and one more likely to be found in moist habitats. Its leaves are pinnately cleft or lobed and coarsely toothed. The flowers are about $^1/_8$ inch wide. It can best be separated from all of the preceding species by its pods, which are less than $^1/_4$ inch long. Late spring–fall.

Isatis tinctoria, Dyer's Woad (plate 174), is a widely branched, glaucous plant with showy panicles of $^1/_4$-inch flowers. The cauline leaves are lanceolate, entire, and auriculate-clasping. The $^1/_2$-inch seed pods droop from short stalks. Cultivated for a blue dye-stuff and escaped to roadsides. Spring.

Erysimum cheiranthoides, Wormseed Mustard (plate 175), is sparingly branched if at all. Its leaves are lanceolate, with entire or slightly toothed margins. The flowers are deep yellow, less than $^1/_4$ inch long, and the slender pods stand erect at the ends of spreading pedicels. Fields and waste places. Summer.

Flowers white (also pink or purple in Hesperis and Lunaria)

Capsella bursa-pastoris, Shepherd's Purse (plate 176). *Capsella, Thlaspi* and *Lepidium* constitute a group of weeds which bear minute, insignificant flowers in spring and are much better known by their flat seed pods. In Shepherd's Purse these are obcordate-triangular, widest at the summit.

Thlaspi arvense is called Penny Cress because of its fruits, which are nearly circular, $^1/_4$ inch to $^1/_2$ inch long, and have a deep notch at the apex. There are numerous auriculate-clasping leaves on the stem.

T. *perfoliatum* has smaller obovate pods and fewer than 6 clasping stem leaves. Fields and waste areas. Spring.

Lepidium virginicum, Peppergrass is smooth and has broadly elliptic pods less than $^1/_4$ inch across, only slightly indented at the apex. Its cauline leaves are tapered toward the base, not clasping.

L. *campestre*, known as Cow Cress, is densely short-hairy, and the stem leaves are auriculate-clasping. Both species are common in waste places. Spring–early summer.

Arabidopsis thaliana, Mouse-ear Cress, is small and has mostly basal leaves, spatulate, hairy, and less than 2 inches long; the few cauline leaves are smaller. Its pods are linear, $^3/_8$ inch long or more. Dry soil. Spring.

Draba verna, Whitlow Grass, blooms very early. All of its leaves are in a crowded basal rosette. The tiny flowers are borne on slender, naked stalks and have petals that are unique among our species in being cleft to the middle. The pods are elliptic, about $^1/_4$ inch long, and half as wide. Disturbed ground.

D. *brachycarpa* also flowers early, but it has a number of narrow leaves on its stems. The pods are also elliptic but extremely small. Early spring.

D. *ramosissima* is known as Rocktwist because of its preference for rocky situations and for its spirally twisted seed pod. It is a mat-forming plant with toothed leaves and blooms later in the spring than the preceding two species.

Arabis lyrata, Lyre-leaved Rockcress. Our plants in this genus are woodland species, often favoring rocky places, and flower in late spring and early summer. A. *lyrata* has a rosette of pinnately lobed basal leaves and narrow, mostly entire cauline leaves tapered to the base. The petals are about $^1/_4$ inch long, and the pods slender, up to 1 $^1/_2$ inches long.

A. patens, Spreading Rockcress, is a hairy plant with erect stems and lanceolate, coarsely toothed stem leaves, sessile, and some clasping. The petals are $^3/_8$ inch long and the pods 1 to 1 $^1/_2$ inches long, very slender, straight, and spreading.

A. canadensis, Sicklepod, is slightly hairy below, smooth above. Its cauline leaves are only slightly toothed and usually not clasping. The pods may be as much as 4 inches long, and are $^1/_8$ inch wide, flat, curved, and pendent.

A. laevigata, Smooth Rockcress (plate 177), is completely smooth and glaucous. Its cauline leaves are lanceolate, coarsely toothed, and sagittate-clasping. The small flowers have greenish white petals, and the spreading pods are 2 to 4 inches long but only $^1/_{16}$ inch wide, flat, and curved. Plants with narrower, sessile cauline leaves not sagittate at the base are differentiated as **var. burkii.**

A. glabra, Tower Mustard, is another glaucous plant but differs from all of the preceding species in having its seed pods stiffly erect, appressed to the stem, and overlapping. Its flowers are yellowish white.

Sibara virginica (*Arabis virginica*), is a small plant, densely hairy, especially in the lower portions. Its basal leaves are deeply pinnately cleft, with up to 15 pairs of very narrow lateral segments; the cauline leaves are similar but smaller. The petals are $^1/_8$ inch long, and the pods flat and $^1/_2$ to 1 inch long. Disturbed ground. Early spring.

Cardamine hirsuta, Hairy Bitter Cress (plate 178), is a small, smooth-stemmed plant also blooming in early spring. It has a rosette of numerous basal leaves deeply pinnately lobed with up to 4 pairs of rounded lateral lobes and a larger, almost round terminal leaflet. There are up to 5 cauline leaves with smaller, narrow divisions. The lower petioles are ciliate near their bases. The pods are erect, up to 1 inch. Lawns and waste areas.

C. parviflora, Small-flowered Bitter Cress, is very similar but the basal leaves are few (if any), small, and without cilia. The cauline leaves have small narrow lateral segments, and the terminal one is only slightly larger.

C. pensylvanica, Pennsylvania Bitter Cress (plate 179), is larger than the previous two. A smooth, succulent species, it is found in wet habitats, often in water. Here again the principal leaves are cauline, but they have oblong, decurrent, lateral lobes, and the terminal segment is much larger and obovate in outline. Its flowers are about $^3/_{16}$ inch wide. Spring.

C. clematitis (plate 180), Bitter Cress, is a plant of wet woodland situations. Its leaves are somewhat irregularly divided, with 3 or 5 large, roundish, shallowly lobed segments; sometimes the upper ones are simple. The petioles have expanded bases. Petals are $^1/_4$ to $^3/_8$ inch long, and pods about 1 inch. Spring.

C. rotundifolia, Mountain Water Cress, is a weak, stoloniferous plant with ovate leaves from 1 to 1 $^1/_2$ inches long; the lower ones may have a pair of very small lateral lobes on the petioles. Its petals are $^1/_4$ inch long, the pods up to $^3/_4$ inch. Low, wet woods. Spring.

C. bulbosa, Spring Cress, is a large-flowered ($^3/_8$- to $^1/_2$-inch white petals) species of wet woods and shallow water, blooming from early spring to summer. All of its leaves are simple, the basal ones few, long-petioled and round to cordate, the cauline ones more numerous, sessile, and elliptic or lanceolate.

C. diphylla (*Dentaria diphylla*), Crinkleroot, Toothwort. This and the next two species are familiar spring woodland wildflowers, with palmate leaves as contrasted with the pinnate or simple leaves of our other *Cardamines*. This species has a pair of subopposite cauline leaves each divided into 3 broad, ovate segments, as well as similar basal leaves. The flowers have white petals $^1/_2$ to $^5/_8$ inch long.

In **C. angustata** (*Dentaria heterophylla*), Slender Toothwort, the basal leaves are similar to the preceding. There also is a pair of cauline leaves, but they are divided into 3 narrowly lanceolate segments less than $^3/_8$ inch wide.

C. concatenata (*C. laciniata*, *Dentaria laciniata*), Cut-leaved Toothwort (plate 181), has a whorl of 3 cauline leaves (the basal leaves are usually absent at flowering time) each deeply 3-parted into narrow segments which are themselves deeply lobed or toothed. The petals are ⁵/₈ inch long, white or pink.

Nasturtium officinale, Water Cress (plate 182). Readily recognizable from its foliage if not by its generic name (which also has been appropriated as a common name for the familiar orange garden flower), this is our only exclusively aquatic species in the Mustard Family. Introduced from the Old World, it has become widely naturalized in clear, flowing streams. It has succulent creeping or floating stems which root at the nodes. Its flowers have ¹/₈-inch white petals and bloom in spring to early summer. The fruits are sickle-shaped, less than 1 inch long.

Alliaria petiolata (A. *officinalis*), Garlic Mustard (plate 183), owes its name to the garlic-like odor of its foliage. The leaves have coarse, sharp teeth and vary in general outline from heart-shaped to triangular. The petals are ¹/₄ inch long, and the slender, erect pods are up to 6 inches long. Waste ground. Spring.

Alliaria *came from England, where it was enjoyed as an economical salad green, to America, where it has become an invasive weed in more than a dozen states—but not yet, at least, in the southern mountains.*

Hesperis matronalis, Dame's Rocket (plate 184), is a large plant, frequently escaping from cultivation. It usually has purple flowers but these may vary to lavender, pink, or white; the petals are about ³/₈ inch long. The leaves are simple and sharply toothed. Spring–summer.

Lunaria annua, Honesty (plate 185) is another popular garden plant that occasionally escapes to the wild. Its leaves are opposite (unusual for this family), cordate-based, and triangular, with coarse teeth. The flowers are red-purple, ³/₄ inch wide. It is also called Money Plant because of the silvery septum remaining after the fruit matures. In this species, this is extremely thin and translucent, broadly oval, and up to 2 inches long. It flowers in late spring or early summer.

Sarraceniaceae, Pitcher Plant Family

Our Pitcher Plants are in the genus **Sarracenia.** As with Sundews, it is the leaves of Pitcher Plants that serve to capture their insect prey, but they act as pitfall traps rather than "flypaper." To complete their task, these hollow structures contain water together with protein-digesting enzymes. The flowers are nodding, on naked stalks, and solitary in all but one species and have 5 incurved deciduous petals, 5 persistent sepals, and a large, rather flat, umbrella-shaped style.

Sarracenia purpurea, which is also known as Sidesaddle Flower (plate 186 and plate 187), has decumbent cornucopia-like leaves up to 10 inches long, narrow at the base, but greatly expanded toward the mouth and prolonged into an erect, ruffled hood. They vary from green to red and are strongly purple-veined. The solitary flower has deep crimson petals 2 inches long. Boggy areas. Late spring–early summer.

S. jonesii *(S. rubra* ssp. *jonesii)*, Sweet Pitcher Plant. In this species the pitchers are slender and erect, 18 inches tall or more, bulbously expanded toward the top, with a flap raised above the orifice; they are reddish tan with a fine network of red veins. Each plant may have several flowers, with red-purple petals only 1 inch long. A federally endangered species found along stream banks in the Carolinas. Late spring–early summer.

S. flava, Yellow Pitcher Plant, Trumpets (plate 188). Although it is very much a plant of the coastal plain, *S. flava* may still be found in isolated mountain locations. Its pitchers are narrow and tall (up to 30 inches), greenish yellow with red veins, and have arching hoods. A few erect, sword-shaped "winter leaves" (phyllodia) may also be present. The flowers have bright yellow 3-inch petals. Boggy situations. Spring.

S. oreophila, Green or Mountain Pitcher Plant. Strictly a mountain species with a very limited range, this bears some resemblance to *S. flava,* but the pitchers are pale green and the petals yellowish green. A diagnostic feature is the large number of phyllodia, which are relatively short and spatulate. Stream banks. Spring.

DROSERACEAE, SUNDEW FAMILY

Drosera rotundifolia, Round-leaved Sundew (plate 189). Much of our fascination with the Sundews stems from the realization that these exquisitely delicate plants actually capture living insects. The trapping is done by the leaves, which bear a multitude of glistening, sticky gland-tipped hairs. The flowers are 5-petaled, in a scorpoid raceme on stalks that, in our species, are without either hairs or glands. In *D. rotundifolia*, our most common species, the leaves are in a basal rosette, with almost circular blades that are up to $^1/_2$ inch wide and not quite as long, on slender petioles. The flowers are white or pale pink, about $^1/_4$ inch across. Bogs and wet slopes. Summer–fall.

D. intermedia (plate 190) has 1-inch spatulate leaves on long petioles, and white flowers. **D. capillaris** has ovate leaves less than $^3/_4$ inch long including the petiole and pink flowers. Both of these species are much more rare in the southern mountains, where they are limited to boggy habitats. Summer.

CRASSULACEAE, ORPINE FAMILY

Sedum acre, Stonecrop, Wallpepper. *Sedums* are characterized by thick, succulent leaves and starlike flowers with 5 (sometimes 4) narrow, sharp-pointed petals. Both of our yellow-flowered species are aliens that have escaped from cultivation to open, rocky areas. *S. acre* is a mat-forming plant with ovoid, sharp-pointed leaves under $^1/_4$ inch long, alternate, crowded, and overlapping. The flowers are $^3/_8$ inch wide on erect branches. Late spring–summer.

In **S. sarmentosum** the leaves are lanceolate, up to 1 inch long, and in whorls of 3; the flowers are $^1/_2$ inch across, in flat-topped, 3-forked inflorescences. Spring–summer.

Our most common white-flowered *Sedum* is the native Wild Stonecrop, **S. ternatum** (plate 191). The flowers, which are $^3/_8$ inch across, are borne in 3-branched terminal cymes. The leaves are spatulate and in whorls of 3. Wet, rocky woods. Spring.

Two other white species, much less common in our area, are **S. glaucophyllum** and **S. nevii.** They have alternate leaves—glaucous in the first and green in the second—and $^5/_{16}$-inch flowers in short cymes. Rocky sites. Spring.

Elf Orpine, **S. smallii** (*Diamorpha cymosa, D. smallii*) (plate 192), has minute white flowers only $^3/_{16}$ inch across, but its red stems and leaves form spectacularly colorful, dense mats in shallow depressions on rock outcrops. Spring.

Rock Moss or Lime Stonecrop, **S. pulchellum,** may be 1 foot or more high and has a forked inflorescence of 3 or more crowded cymes. Its flowers are $^3/_8$ inch wide and range from white to rose-purple. The leaves are linear, almost round in cross-section, and up to 1 inch long. Spring.

S. telephioides (plate 193) has numerous common names, among them Wild Live-forever, Cliff Orpine, and Allegheny Stonecrop. It is our tallest *Sedum* (up to 18 inches) and has well-spaced alternate, obovate leaves 1 to 2 inches long, and a rounded terminal cluster of pink flowers about 3 inches across. The foliage is purple-tinged and has a whitish bloom. Rocky woods and cliffs. Summer–fall.

Roseroot, **S. rosea,** is not quite so tall. It has green, crowded, sessile leaves 1 to 1 $^1/_2$ inches long, and a compact terminal inflorescence about 1 inch across. Its flowers are unisexual, the staminate ones greenish yellow and the pistillate purplish red. Open cliffs at high elevations. Spring–summer.

SAXIFRAGACEAE, SAXIFRAGE FAMILY

Saxifraga michauxii, Michaux's Saxifrage (plate 194 and plate 195). Our Saxifrages have simple, toothed basal leaves, and only small bractlike leaves, if any, in the branched inflorescences. The flowers have 5 narrow petals and 10 stamens. S. *michauxii* is 1 to 1 $^1/_2$ feet high and is the only one with bilaterally symmetric flowers— the 3 upper petals clawed and auriculate, and with yellow spots, the 2 lower ones spatulate and unspotted— the whole $^3/_8$ inch across. Its leaves are up to 6 inches long, obovate, and very coarsely toothed. Moist rocks, slopes, and balds. Summer.

S. micranthidifolia, Mountain or Brook Lettuce (plate 196). This is a taller (up to 3-foot) plant with oblanceolate, serrate leaves long-attenuate at the base, often 8 inches or longer. Its flowers are white, radially symmetric, and less than ¹/₄ inch wide. Streams and seeps. Spring–early summer.

S. virginiensis, Early Saxifrage (plate 197). Usually under 1 foot in height, this is one of our early spring bloomers. It has a basal rosette of ovate leaves about 2 inches long, tapering gradually at the base, and usually toothed. The flowering stem is thick and downy, and bears a comparatively compact cymose cluster of ³/₈-inch flowers with white spatulate petals. Dry, rocky banks and fields.

S. careyana grows to about 1 ¹/₂ feet high and has ovate, toothed leaves under 4 inches long, abruptly narrowed to a winged petiole. The flowers, borne in an open, widely spreading panicle, are white and about ¹/₄ inch wide. Confusingly similar is **S. caroliniana,** a close relative if indeed it is a distinct species. The key characters tend to be inconsistent, but among those attributed to *S. caroliniana* are ovate (instead of oblong or elliptic) petals, yellow spots, and clavate (rather than filiform) filaments. Both inhabit moist, rocky places and flower in late spring and early summer.

S. pensylvanica, Swamp Saxifrage (plate 198), is a hairy plant with a very stout flowering stem sometimes arising to a height of 4 feet from a rosette of 4- to 8-inch lanceolate leaves which may be entire or obscurely toothed. The inflorescence is compact and the flowers only ³/₁₆ inch across; the petals are narrow and may be white but more often are tinged with green, yellow, or purple. Swamps and wet meadows. Spring–early summer.

The ability of Saxifrages to take root in small crevices gives the appearance of their having broken the rocks among which they grow. In fact, the name comes from the Latin words saxum *(rock) and* frangere *(to break). According to the doctrine of signatures, therefore, it was thought they should provide an effective remedy for kidney stones.*

Heuchera villosa, Alumroot (plate 199 and plate 200). Alumroots have long-stalked, palmately lobed, toothed basal leaves and panicles of very small flowers with 5 stamens. This and the next species bloom from mid-summer to early fall, the others in spring and early summer. (The latter can be distinguished from Foamflower, *Tiarella,* by the multiple racemes and the 5 rather than 10 stamens.) In *H. villosa* both the lobes and teeth of the leaf blades are sharp-pointed. The petioles often have rusty hairs, and the flowering stem, leaf undersides, and calyxes are soft-hairy. The flowers are white or pink, and the stamens protrude beyond the style. Rocky cliffs and woods.

H. parviflora has almost circular leaves, and the lobes and teeth are broadly rounded; both sides are soft-hairy, as are the petioles, stem, and calyxes. Its flowers are whitish, and the stamens do not surpass the style. Shaded rocks and ledges.

The leaves of *H. americana* are shallowly and obscurely 5-lobed with mostly blunt teeth. The plant is generally smooth but has some short glandular hairs in the tall, slender inflorescence. The flowers are pinkish or purplish, and both the stamens and the style are conspicuously exserted. Dry woods.

In two very similar species, the calyxes are irregular (the upper lobes longer than the lower), the sepals are about as long as the petals, and the stamens are not exserted. In *H. pubescens* the style is conspicuously longer than the petals; in *H. longiflora* it is included.

Mitella diphylla, Miterwort, Bishop's Cap (plate 201 and plate 202). This species is named for the pair of horizontal, sessile leaves near the middle of the stem. Its tiny white flowers are borne in a long, slender spike-like raceme; they resemble snowflakes, and must be seen with a hand lens to be fully appreciated. Less than 1/4 inch in diameter, they have a cuplike hypanthium and 5 spreading petals that are deeply pinnatifid into thread-like segments. Rich woods. Spring.

Tiarella cordifolia (*T. wherryi*), Foamflower, False Miterwort (plate 203). The name False Miterwort sometimes causes confusion with *Mitella,* but *Tiarella* has no cauline leaves and its petals are not pinnatifid. It is about 1 foot high, with a raceme of white flowers on 5/8-inch pedicels. Each has 5 white clawed, elliptic petals and 10 conspicuously long stamens with apricot-colored anthers. Rich woods. Spring–early summer.

Boykinia aconitifolia, Brook Saxifrage, has long-stalked basal leaves up to 5 inches wide, palmately lobed and sharp-toothed, and small sessile or short-stalked cauline leaves. The inflorescence is open, on a stalk as much as 2 feet high. The $^{1}/_{4}$-inch, white 5-petaled flowers have 5 stamens, unlike the true Saxifrages, which have 10. Wet woods, cliffs, and stream banks. Summer.

Astilbe biternata, False Goatsbeard (plate 204). As the common name suggests, this species of *Astilbe* closely resembles *Aruncus dioicus* in the Rose Family (one of many plants familiarly known as Goatsbeard), but it has a hairy upper stem whereas *Aruncus* is smooth, and its terminal leaflets are usually 3-lobed. Further, the flowers of *Astilbe* may be either perfect or unisexual, and the pistils are 2, the stamens 10. Rich woods. Early summer.

Parnassia asarifolia, Grass-of-Parnassus (plate 205). Our species of *Parnassia* have several long-petioled entire basal leaves and a single, smaller leaf clasping the stem near the middle. Their solitary flowers would be attractive enough, with their 5 white petals delicately veined in green, but they are further enhanced by 5 staminodia, each of which consists of 3 sterile stamens joined at the base. *P. asarifolia* has kidney-shaped leaves 1 to 2 inches long and somewhat wider. Its $^{1}/_{2}$-inch petals are narrowed to a claw and each has 11 to 15 principal radiating veins. The staminodia are $^{3}/_{8}$ inch long, about equaling the fertile stamens. Rocky seepage slopes. Late summer–fall.

P. grandifolia is a little larger, with ovate leaves 2 to 3 inches long and not quite as wide. The petals are $^{3}/_{4}$ inch long, not clawed, and have fewer veins than *P. asarifolia*. The staminodia are slender, often exceeding the stamens in length. Seepage areas. Fall.

Penthorum sedoides, Ditch Stonecrop (plate 206). In some works this will be found with other Stonecrops and Sedums in the Orpine Family, but it differs from them in not being succulent. Ditch Stonecrop grows up to 2 feet tall, with alternate, narrow, lanceolate leaves with finely toothed margins, and terminal cymes 1 to 3 inches long. These bear secund rows of greenish flowers with 5 pointed sepals (no petals) and 10 stamens. Wet areas and muddy soil. Summer–fall.

Chrysosplenium americanum, Golden Saxifrage, Water Mat (plate 207). This is an insignificant creeping plant growing in springs, seeps, and small brooks, and its unusual little flowers are often overlooked. They are only ³/₁₆ inch wide, with 4 or 5 spreading green or yellowish sepals, and several short stamens growing from a ring-shaped reddish disk. The leaves are ovate and under ¹/₂ inch long. Spring.

ROSACEAE, ROSE FAMILY

***Geum* spp.,** Avens. *Geum* is a group of plants with variable foliage but usually having some leaves pinnately divided or lobed, the terminal segment larger than the lateral ones. The few flowers have 5 petals, 5 sepals, and numerous stamens and pistils. In all but one species the persistent achenes are tipped by a hooked style.

In the common White Avens, ***Geum canadense*** (plate 208), the lower leaves are divided into 3 or 5 obovate leaflets. The stems are smooth or at most have some appressed hairs. Its flowers are ¹/₂ inch across, the white petals about as long as the sepals. Woodlands. Late spring–summer.

Rough Avens, **G. *laciniatum,*** has a hairy stem and leaves that are deeply lobed or dissected. The white petals are much shorter than the sepals. Found in wet meadows during the summer, it is less frequent than G. *canadense*.

Bent Avens, **G. *geniculatum,*** has leaves with 3 lobes or leaflets. It bears a panicle of white or pinkish flowers with spatulate petals about 1/4 inch long and pistils ³/₈ inch long. It grows only at high elevations and is uncommon.

Most of the leaves of Spring Avens, **G. *vernum,*** are dissected into pinnate segments that are themselves deeply incised. It bears ¼-inch, yellowish flowers; the petals are nearly as short as the sepals, which are reflexed. The presence of a stalk between the calyx and the head of the achene is diagnostic. Moist woods.

G. virginianum is densely soft-hairy. The principal leaves are irregularly pinnate, with mostly ovate to lanceolate segments. Its flowers are creamy yellow, and its ¹⁄₁₆-inch petals are exceeded by the sepals. Woods. Summer.

In Yellow Avens, **G. aleppicum,** the leaves are irregularly lobed; the segments are distinctly wedge-shaped at their bases and usually are accompanied by several very small leaflets. The flowers are yellow, about ³⁄₄ inch wide. Wet places. Summer.

Spreading Avens, **G. radiatum** (plate 209), has very large basal leaves, the terminal lobe kidney-shaped and up to 6 inches wide (and the petioles even longer), but the lateral ones less than 1/2 inch. The flowers are bright yellow, 1 ¼ inch across. Its achenes have a hairy style, not hooked. Rocky balds at high elevations. Summer.

Agrimonia spp., Agrimony. Agrimony foliage resembles that of *Geum*—although the terminal leaflets are only slightly if at all larger than the others—but its flowers are disposed along slender, wandlike racemes. They are 5-petaled, yellow, and mostly ¹⁄₄ inch or less in diameter, and the hypanthium has several rows of hooked bristles.

Agrimonia parviflora (plate 210) has 11 or more principal leaflets (the other species have fewer). Along with the next three species, it may grow as tall as 6 feet. Thickets. Summer–early fall.

A. pubescens can be distinguished by its densely pubescent stem and the velvety undersides of its leaves. Woodlands. Summer–fall.

In **A. gyrosepala** the leaf undersides are minutely glandular, and the leaflets are elliptic and pointed at the apex. Its stipules are more than ³⁄₈ inch wide and foliaceous. **A. rostellata** has rounded or oblanceolate leaflets with blunt teeth, and the stipules are under ¹⁄₈ inch.

Potentilla spp., Cinquefoil. The name Cinquefoil denotes 5 leaves (or leaflets), but this holds true only for the first three of our species. Otherwise, they all have in common yellow flowers with 5 roundish petals, 5 sepals interspersed with 5 bracts (the presence of these bracts helps to distinguish them from species of *Ranunculus*), and numerous stamens and pistils. They have adapted to a variety of habitats, most of them open and sunny.

Potentilla simplex, Common Cinquefoil (plate 211), is prostrate and has leaves palmately divided into 5 elliptic leaflets with teeth that usually extend below the middle. The flowers are $^1/_2$ inch wide, solitary, and axillary, the first flower appearing in the axil of the second fully developed leaf. Spring.

Dwarf Cinquefoil, **P. canadensis** (plate 212), is similar, but the leaflets are obovate, more wedge-shaped at the base, and for the most part are toothed only above the middle. The first flower arises from the axil of the first mature leaf. Spring.

P. argentea, or Silvery Cinquefoil, also has 5 leaflets, but they are narrow with a few linear toothlike lobes, and are densely white-woolly beneath. The flowers are about $^3/_8$ inch wide and, together with the succeeding species, are cymose. Spring–summer.

Rough Cinquefoil, **P. norvegica** (plate 213), is a stout plant up to 3 feet high, with coarsely toothed trifoliolate leaves. Its flowers are up to $^3/_8$ inch across. Spring to fall.

P. recta, known as Rough-fruited or Sulphur Cinquefoil (plate 214), is the largest of our species, and has flowers up to 1 inch wide with pale yellow petals notched at the apex, in a flattened inflorescence. Its leaves have 5 to 7 narrow leaflets, toothed throughout. Spring–summer.

Sibbaldiopsis tridentata (formerly named *Potentilla tridentata* and still known familiarly as Wine-leaved Cinquefoil) (plate 215). This has white flowers and trifoliolate leaves, dark green but turning crimson in the fall; the leaflets are wedge-shaped and have 3 teeth at the apex. The flowers are between $^1/_4$ and $^1/_2$ inch wide.

It is found on balds and rock outcrops, only at high elevations in our area. Summer.

For the most part, the range of Wine-leaved Cinquefoil extends from Greenland to southern Canada and northern United States, where it is often found at sea level. In warmer latitudes such as ours it is a disjunct, able to survive only in scattered locations at high altitudes where the climate is suitably cold.

Fragaria virginiana, Wild Strawberry (plate 216). Both of our Wild Strawberry species are stoloniferous plants with trifoliolate, toothed leaves and small clusters of white, 5-petaled flowers on separate stalks. In this, the more common, the leaves often overtop the flowers, which are between $1/2$ and 1 inch wide. There are 5 sepals alternating with 5 similar but smaller bracts. The fruits are red, ovoid, and juicy, with the "seeds" (achenes) embedded in pits. Fields, clearings, and woodland margins. Spring–summer.

F. vesca ssp. americana, known as Wood Strawberry, prefers cool, moist woods and is much more scarce in our region. Its flower stalks often surpass the leaves, and the blossoms are only about $1/2$ inch across. The fruits are more tapered, and the achenes not embedded. Spring–summer.

Indian Strawberry, **Duchesnea indica** (plate 217), is a trailing plant with foliage resembling that of *Fragaria* and solitary 5-petaled flowers on long stalks from the axils. The blossoms, however, have yellow cuneate petals and are $1/2$ to $3/4$ inch wide. The sepals alternate with large, leafy bracts which have 3 prominent teeth. The fruits look like strawberries but are dry. Waste places and lawns. Spring-summer.

Also with trifoliolate leaves and yellow blossoms, Barren Strawberry, **Waldsteinia fragarioides** (*W. doniana, W. parviflora*) (plate 218), bears a superficial resemblance to *Duchesnea*, but the leaflets are more round and irregularly lobed. The flowers are from $1/4$ to $1/2$ inch wide and are borne in cymes, and the conspicuous bracts are lacking. **W. lobata** is a rare species with leaves that are 3-lobed (not divided) and narrow petals. Both grow in wooded areas and bloom in spring.

Dalibarda repens, Dewdrop (plate 219), is our only herbaceous member of the Rose Family with simple, unlobed leaves; they are cordate-based, almost circular, with scalloped margins, and arise from a creeping stem along with flowers of two types. The $1/2$-inch, sterile flowers are on reddish 2- to 4-inch scapes and have 5 white, elliptic petals and many long, slender white stamens; the fertile ones are on shorter peduncles and lack petals. Moist woods. Summer.

Rubus odoratus, Purple-flowering Raspberry (plate 220). Traditionally, most plants in the genus *Rubus*—collectively referred to as brambles—are excluded from popular wildflower guides on the basis of their woody character. This species usually gains admittance, however, by virtue of at least two unique features that make it visually appealing. For one thing, its large leaves, being simple and merely palmately lobed rather than divided into separate leaflets, are eye-catching even at a distance. Then there are its handsome blossoms, which are larger (1 to 2 inches) and more colorful (rose-purple) than those of its fellows. As a bonus, it has glandular hairs but never thorns. Rocky woods and forest margins. Summer.

Porteranthus trifoliatus *(Gillenia trifoliata),* Bowman's Root, Indian Physic, Fawn's Breath (plate 221). An open-branching habit and long, slender flower stalks give these plants a delicate, airy aspect. In this species the principal leaves are ternately compound and irregularly toothed, and the stipules are linear and inconspicuous, only about $1/4$ inch long. The flowers have 5 narrow $1/2$-inch, white petals, spreading and slightly twisted. The reddish hypanthium sometimes causes the flowers to appear pinkish at a distance. Woods and road banks. Spring.

American Ipecac, ***P. stipulatus*** *(Gillenia stipulata),* is similar except for its stipules, which are ovate, variously toothed or lobed, and $3/4$ inch or more long; because of their leaflike appearance they sometimes are mistaken for additional leaflets. Some of its lower leaves may have pinnately lobed or divided leaflets. Often associated with basic soils. Spring.

Filipendula rubra, Queen-of-the-Prairie (plate 222), is a very showy plant, up to 8 feet high with a large spreading panicle of deep pink 5-petaled flowers $^3/_8$ inch wide. The lower leaves are pinnately compound, but the terminal leaflet is very large, up to 8 inches wide, and is deeply cleft into irregularly toothed lobes; the lateral leaflets are much smaller but also lobed. Moist meadows. Summer.

Aruncus dioicus, Goatsbeard (plate 223). Several very different plants are known by the name Goatsbeard. This one grows to a height of 6 feet or more and is erect but often leans out from road banks at the edges of woods. Its leaves are 2 or 3 times ternately compound with long-pointed, ovate leaflets. The inflorescence is a branching cluster of long, narrow spikes of tiny creamy white 5-merous flowers. The pistillate flowers have 3 or 4 carpels with divergent styles, and the staminate flowers—which are on separate plants—have 15 or more stamens. (See plate 204, *Astilbe biternata*, in the Saxifrage Family, for a comparison of these remarkably similar species.) Spring–early summer.

Sanguisorba canadensis, American or Canadian Burnet (plate 224). The lower leaves of this plant are pinnately divided into 7 or more toothed, oblong segments. The flowers are white, in dense cylindrical spikes up to 6 inches long, and reaching a height of 6 feet above the ground. There are 4 tiny petaloid sepals but no petals; the fuzzy appearance of the inflorescence is due to the white filaments, which are up to $^1/_2$ inch long. Seeps, bogs, and wet meadows. Summer–fall.

MIMOSACEAE, MIMOSA FAMILY

Schrankia microphylla, Sensitive Brier (plate 225). Although *Schrankia* is frequently placed in the Bean Family (Fabaceae), it is separated here because of its markedly different flowers. *S. microphylla* is a sprawling herb, prickly with hooked thorns. Its leaves are bipinnately compound; the ultimate segments are very numerous, oblong, less than $^1/_4$ inch long, and fold when touched. The flowers are numerous in $^3/_4$-inch spherical heads, each having a very small radially symmetric, funnelform, 5-lobed corolla and 10 long, slender, rose-pink stamens. Dry soil. Summer.

Our "sensitive plants" are interesting as curiosities, but their reaction to distur-bance is disappointingly slow. The name is better deserved by the tropical Mimosa pudica *whose leaves fold almost instantaneously—much like those of Venus' Flytrap. The reason behind this apparently defensive mechanism is the subject of much speculation.*

CAESALPINIACEAE, CASSIA FAMILY

Our flowers in this family have 5 yellow petals radially arranged but unequal in size, the lowermost the largest. Our species are herbs with even-pinnately compound leaves. The shape of a small gland on the petiole is some-times useful in identifying species.

Chamaecrista fasciculata (*Cassia fasciculata*), Partridge Pea (plate 226), is a common, often abundant plant blooming on roadsides and in waste places during the summer. It grows to 2 feet high and has numerous nar-row, oblong leaflets $^1/_2$ to $^3/_4$ inch long. The flowers are solitary or few in the leaf axils, 1 inch wide, the petals broad, some with a red blotch at the base. There are 8 short stamens with yellow anthers and 2 longer ones with purple anthers.

C. nictitans (*Cassia nictitans*), Wild Sensitive Plant (plate 227), is much smaller, the leaflets under $^1/_2$ inch and sensitive to the touch, and the flowers $^1/_4$ inch across. The lowermost petal is about twice as long as the others. There are only 5 stamens, unequal but all perfect. Dry soil. Summer–fall.

Senna hebecarpa (*Cassia hebecarpa*), Wild Senna, has 5 to 10 pairs of leaflets, elliptic, 1 to 2 inches long. The flowers are $^3/_4$ inch wide and numerous in axillary racemes. The upper 3 stamens are sterile, the lower 7 fertile. The petiolar gland is club-shaped, on a short stalk. **S. marilandica** (*Cassia marilandica*) is similar but has fewer flowers, and the petiolar gland is dome-shaped, not stalked. Moist, open areas. Summer.

S. obtusifolia (*Cassia obtusifolia*), Sicklepod, has only 2 or 3 pairs of leaflets; they are obovate and widely rounded at the summit, the terminal ones largest and up to 2 $^1/_2$ inches long. The flowers are up to $^3/_4$ inch wide, usually solitary in the upper axils. There are 10 stamens, some of them sterile. Moist woods and waste places. Summer–fall.

Fabaceae, Bean Family

The 5-parted flowers in the Bean Family are of a unique form known as "papilionaceous." The upper petal, called the standard, is usually the largest and stands more or less erect; the pair of lateral petals are known as the wings, and the two lowermost ones are partially fused into a so-called keel. This last encloses the stamens and style.

The seeds in this family (and in the two preceding ones as well) are contained in pods that usually split open along longitudinal seams and are called "legumes," a term also applied to any plants that bear such fruits. In some genera, however, the pods are constricted into segments each bearing a single seed, in which case they are referred to as "loments."

The leaves are alternate and (with a single exception) are compound. In a few genera there are palmate leaves (that is, with the terminal leaflet sessile), but in most they are pinnate (with the terminal leaflet stalked). Among the latter the terminal leaflet is sometimes replaced by a tendril.

Leaves simple: *Crotalaria*
Leaves with tendrils: *Lathyrus*, *Vicia*
Leaflets 3; stems twining: *Clitoria*, *Centrosema*, *Pueraria*,
   Phaseolus, *Strophostyles*, *Amphicarpaea*, *Galactia*
Leaflets 3; stems not twining: *Baptisia*, *Thermopsis*,
   Stylosanthes, *Rhynchosia*, *Psoralea*, *Desmodium*, *Lespedeza*,
   Trifolium, *Medicago*, *Melilotus*
Leaflets more than 3: *Apios*, *Lotus*, *Coronilla*, *Tephrosia*,
   Astragalus, *Lupinus*

Leaves simple

Crotalaria sagittalis, Rattlebox, is the only member of the Fabaceae in our area with undivided leaves. It is a hairy plant 1 foot or more high, with lanceolate to linear leaves. The flowers are yellow, about $1/2$ inch long, in few-flowered racemes. The specific epithet refers to the conspicuous inverted arrowhead-shaped stipules. Dry, open woods and waste areas. Summer.

Leaves with tendrils

Lathyrus latifolius, Everlasting Pea, has broadly winged stems and petioles, leaves consisting of a single pair of lanceolate leaflets up to 3 inches long, and a branched terminal tendril. The flowers are $^3/_4$ to 1 inch long, purplish pink, as many as 10 in long-stalked axillary racemes. (The familiar annual Sweet Pea, *L. odoratus,* is similar but has very few flowers to a raceme.) Escaped to waste areas. Late spring and summer.

L. hirsutus is more narrowly winged and its leaflets are less than 2 inches long. The flowers are violet to purple and $^1/_2$ inch long, one or a few in each cluster. Roadsides and waste areas. Spring–summer.

L. venosus, Veiny Pea (plate 228), has a 4-angled (but not winged) stem and 4 to 6 pairs of elliptic leaflets 1 to 2 inches long, terminated by a branched tendril. The inflorescences are dense racemes of 10 to many more purple flowers $^1/_2$ to $^3/_4$ inch long. Thickets and stream banks. Spring–summer.

Vicia spp., Vetch. The Vetches are slender, sprawling plants with their terminal leaflets modified into a tendril. As a rule, they have smaller and more numerous leaflets than *Lathyrus*.

Vicia grandiflora, Yellow Vetch, has large (up to 1 $^1/_2$-inch) yellow flowers, usually 2 nearly sessile in each upper axil. There are from 6 to 12 leaflets. Open areas. Spring–early summer.

V. sativa, Spring Vetch, also has 2 flowers on very short stalks in the axils, but they are smaller (about 1 inch) and red-purple with darker violet wings. There are 8 to 16 oblong to elliptic leaflets from $^3/_4$ to 1 $^1/_4$ inches long. Fields and roadsides.

V. angustifolia, Narrow-leaved Vetch (plate 229), has pairs of even smaller (about $^3/_4$-inch), rose-purple flowers on short stalks. Its 6 to 12 leaflets are very narrow and under 1 inch long. Fields and waste areas. Spring–summer.

V. americana, Purple Vetch. This species (and the remaining ones) has its flowers in peduncled racemes; in this case there are 9 or fewer ¹/₂- to ³/₄-inch, bluish purple flowers in each. Moist woods and meadows. Late spring–summer.

V. cracca, Cow Vetch, has long-stalked, one-sided racemes densely crowded with numerous blue-purple flowers that are ¹/₂ inch long and hang downward. Its leaves have 5 to 12 pairs of narrowly elliptic leaflets. Roadsides and fields. Late spring–summer.

V. dasycarpa, Smooth Vetch (plate 230), may have from 10 to 30 flowers in a raceme. There are 10 to 20 leaflets. **V. villosa,** Hairy Vetch, is similar except for being villous throughout. In both species the flowers are bicolored—violet and white—and the flower stalk is attached to the underside of the calyx. Open areas. Spring–summer.

V. caroliniana, Wood Vetch (plate 231), is a delicate woodland species blooming in spring. Its flowers are ³/₈ inch long, white with a blue-tipped keel.

Leaflets 3; stems twining

Clitoria mariana, Butterfly Pea (plate 232), is a twining herbaceous vine with large flowers occurring singly (or very few) on stalks from the leaf axils. They are pinkish lavender with a standard up to 2 inches high, obovate in general outline, the wings and keel much smaller. The leaves of *Clitoria* and all the other twining plants in the Bean Family are pinnately compound. Open areas. Summer.

Centrosema virginianum, Spurred Butterfly Pea (plate 233), is a more slender plant, and the standard is rounder and smaller, little more than 1 inch across. The calyx has a short tube and longer lobes, whereas the reverse is true in *Clitoria*. Open areas. Summer.

Both Centrosema *and the preceding genus,* Clitoria, *may have flowers with the curious habit of turning around and growing upside down, with the keel uppermost.*

Pueraria lobata, Kudzu (plate 234 and plate 235). Our only other twining vine with large flowers is the semi-woody Kudzu. This deep-rooted, fast-growing, nitrogen-fixing plant, a native of Japan, seemed ideal for limiting soil erosion and enriching farmland in the United States, but in the absence of natural biological constraints here, it has spread out of control in many parts of the South, completely covering trees, utility poles, and even abandoned vehicles and buildings. Its leaves are trifoliolate, the leaflets variable in shape, sometimes lobed, and as much as 8 inches long. The flowers are violet to purple, in dense clusters, and have a grapelike aroma; the standard is about 1 inch high, with a yellow basal blotch. Summer–fall.

Phaseolus polystachios, Wild Bean (plate 236), is a twining vine with ovate leaflets, the lateral ones often asymmetric. The flowers are purple, $^3/_8$ inch long, with the keel petals spirally coiled. They are borne in slender, long-stalked, loosely flowered racemes. Thickets. Summer.

P. sinuatus is a trailing vine and its leaflets often are shallowly 3-lobed. The flowers are pale purple and slightly smaller.

Strophostyles umbellata, Pink Wild Bean. In this genus the flowers are tightly clustered in very short racemes at the end of long, rigid axillary stalks. They are purplish pink, about $^1/_2$ inch long, with an almost round standard, and the keel is strongly curved upward (but not coiled), forming a beak. Dry, open areas. Summer.

S. helvula, Trailing Wild Bean (plate 237), has broader leaflets than the preceding, and they frequently have rounded lobes.

Amphicarpaea bracteata, Hog Peanut (plate 238), is a very slender, delicate climber with thin, ovate leaflets. The flowers are pale purple to almost white, about $^1/_2$ inch long, the keel nearly straight. They are crowded in stalked axillary clusters. Woods, thickets, and roadsides. Summer.

Galactia regularis, Milk Pea (plate 239), has only 1 to 3 flowers in each axil; they are pink, about $^1/_2$ inch long, with straight keel petals. The leaflets are oblong to elliptic, rounded at the apex. Open areas. Summer. This is a prostrate plant that sometimes twines at the top; it is often confused with **G. *volubilis***, a vigorous climber with slightly smaller flowers.

Leaflets 3; stems not twining

At one time **Baptisia tinctoria,** known as Wild Indigo (plate 240), served as a substitute for true indigo (*Indigofera*), which is grown in warmer parts of the world. Only two genera in the Bean Family—*Baptisia* and *Thermopsis*—have entire, palmately trifoliolate leaves. In this species they have small (under $^3/_4$-inch) obovate leaflets tapered at the base and nearly sessile. This is a widely branched plant up to 3 feet high, smooth and glaucous, turning black when dry. The flowers are yellow, $^1/_2$ inch long, few at the ends of branches. Dry woods and clearings. Spring–fall.

B. bracteata (B. *leucophaea*), Cream Wild Indigo, is a hairy plant about 2 feet tall, usually with a single declined one-sided raceme. The flowers are creamy white, 1 inch long, their stalks subtended by persistent foliaceous bracts more than $^1/_2$ inch long. The leaflets are oblanceolate, up to 1 $^1/_2$ inch long. Dry, open woods. Spring.

B. alba (B. *albescens*), White Wild Indigo (plate 241), is smooth and glaucous, with leaflets about 1 $^1/_4$-inch-long and $^5/_8$-inch white flowers in one or a few erect racemes; the floral bracts are narrow, less than $^1/_4$ inch long, and fall early. Dry areas. Spring.

B. australis, Blue False Indigo, is another glaucous species and may grow as tall as 5 feet. Its leaflets are from 1 to 3 inches long, and the flowers are large ($^3/_4$ to 1 $^1/_4$ inches) and blue or violet. Moist woods and thickets. Spring–early summer.

Thermopsis villosa, Hairy Bush Pea (plate 242). This genus could easily be mistaken for *Baptisia*, but the flowers are yellow and the leaflets from 1 to 4 inches long, whereas our only yellow-flowered *Baptisia* has much smaller leaves. Also, there are conspicuous and persistent stipules in *Thermopsis*. In this species these stipules may be 2 inches long; they are ovate and have clasping basal lobes. Open woods and clearings. Late spring–early summer.

Two other species have narrow stipules that do not clasp the stem, and have fewer flowers. **T. mollis** is downy and the floral bracts are longer than the individual flower stalks (pedicels). **T. fraxinifolia** is a smooth plant and the bracts are shorter than the pedicels.

Stylosanthes biflora, Pencil Flower (plate 243), is a wiry plant with narrowly lanceolate or elliptic, bristle-tipped leaflets about 1 inch long. The flowers are bright orange-yellow with an almost circular standard $^{1}/_{4}$ inch high, solitary or few in very short spikes at the ends of the stems. Dry soil in fields and waste places. Summer.

Rhynchosia tomentosa is an erect, hairy plant. Its leaves have ovate leaflets from 1 $^{1}/_{4}$ to 2 $^{1}/_{4}$ inches long, the undersides densely gray-woolly. The flowers are yellow, about $^{1}/_{4}$ inch long, and are crowded in axillary racemes. Dry woods and roadsides. Summer.

Psoralea psoralioides var. eglandulosa, Sampson's Snakeroot (plate 244), has leaflets 1 to 3 inches long but one-fourth as wide or less. The $^{1}/_{4}$-inch flowers are blue-purple, in long slender racemes that usually rise well above the leaves. Open woods and fields. Late spring–summer.

P. onobrychis, Sainfoin, is a less common woodland species, with ovate leaflets up to 4 inches long and half as wide. The racemes are sometimes overtopped by the foliage. Summer–fall.

Desmodium spp., Tick Trefoil, Sticktight. The common names given to the *Desmodiums* refer to their fruits, which are covered with minute hooked hairs (see *D. nudiflorum,* plate 247). These pods (called loments) are flat strips of 2 or more triangular or oval segments, each containing a seed; they serve to differentiate between this and the next genus, *Lespedeza,* and are sometimes useful in distinguishing among species of *Desmodium.* All of our species have purplish flowers less than $^{1}/_{2}$ inch long and bloom during the summer. They are most commonly found in open woods or fields or along roadsides. In addition to those described below, there are a number of disjunct or peripheral species that occur in our area very infrequently.

Leaflets round: *D. rotundifolium*
Leaflets narrow: *D. paniculatum*
Leaflets ovate, leaves clustered: *D. nudiflorum, D. glutinosum*
Leaflets generally ovate, leaves alternate; loments with 1 to 3
  rounded segments; flowers under ¹/₄ inch: *D. obtusum,*
  D. nuttallii, D. ciliare, D. marilandicum
Leaflets generally ovate, leaves alternate; loments with 4 or
  more angular segments; flowers over ¹/₄ inch: *D. canescens,*
  D. cuspidatum, D. laevigatum, D. viridiflorum, D. perplexum

Desmodium rotundifolium, Prostrate Tick Trefoil, Dollarleaf (plate 245), is unique in that its leaflets are nearly round, usually 1 to 2 inches across. It is a hairy plant with erect, sparsely flowered racemes arising from the trailing stem.

D. paniculatum, Panicled Tick Trefoil (plate 246), may be singled out as the only one with narrow leaflets, averaging one-fourth as wide as long. They vary in shape from lanceolate to linear, and have long petioles. The inflorescence is a large, widely branched panicle.

D. nudiflorum, Naked-flowered Tick Trefoil (plate 247). This and the next species have their leaves crowded in a whorl-like cluster on an erect stem. In *D. nudiflorum* this sterile stem is less than 1 foot high; the flowers are borne on a leafless stem up to 3 feet long, which arises separately from the base of the plant.

D. glutinosum, Pointed-leaved Tick Trefoil. In this species there is only one stem, and it is prolonged above the clustered leaves into an elongate terminal panicle of flowers.

In **D. obtusum** (*D. rigidum*) the leaflets are more than 1 inch long and less than half as wide, and essentially smooth beneath; the stipules fall early. **D. Nuttallii** differs chiefly in having persistent stipules and leaves which are velvet-hairy beneath.

D. ciliare, Small-leaved Tick Trefoil, has small, blunt leaflets (usually 1 inch or less in length and more than half as wide), with petioles under ³/₈ inch, and is more or less soft-hairy. **D. marilandicum** is similar but smooth and has long petioles.

D. canescens, Hoary Tick Trefoil, is a much-branched, finely hairy plant with leaflets up to 4 inches or more in length. It has prominent stipules, ovate and up to $^1/_2$ inch long.

D. cuspidatum, Long-bracted Tick Trefoil, is a mostly smooth plant with large ovate leaves, the terminal leaflet tapering to a long point, the stipules lanceolate. The inflorescence is a loose panicle.

D. laevigatum, Smooth Tick Trefoil, is smooth-stemmed and its leaves are glaucous beneath. **D. viridiflorum,** on the other hand, has finely hairy stems, and the leaf undersides are densely velvet-hairy.

D. perplexum (*D. paniculatum* var. *dillenii, D. dillenii, D. glabellum*), Dillen's Tick Trefoil. Judging from the number of synonyms, the authorities have had difficulty agreeing on a name for these variable plants, but at least they differ from the typical *D. paniculatum* in having ovate leaflets about half as wide as long. They are appressed-hairy on both surfaces, paler beneath.

Lespedeza spp., Bush Clover. Bush Clovers are to be found along roadsides and in fields and waste places (where they flower in summer and fall) mostly as a result of their having been planted widely to replenish nitrogen in the soil and to control erosion. They may be distinguished from *Desmodium* by their fruits, which are single and one-seeded and are without hooked bristles. Our species can conveniently be divided according to the principal color of their flowers, which are less than $^1/_2$ inch long.

> Flowers purplish; stems reclining: *L. repens, L. procumbens, L. stipulacea, L. striata*
> Flowers purplish; stems erect: *L. violacea, L. virginica, L. intermedia, L. nuttallii, L. stuevei*
> Flowers yellowish white: *L. capitata, L. hirta, L. cuneata*

Lespedeza repens, Creeping Bush Clover (plate 248), has a prostrate stem, smooth or appressed-hairy, with ascending flowering branches. The leaflets are oval, about $^1/_2$ inch long. The petals are equal in length.

L. procumbens, Trailing Bush Clover, is similar but has spreading hairs.

L. stipulacea *(Kummerowia stipulacea)*, Korean Clover, and **L. striata** *(K. striata)*, Japanese Clover, both Asian introductions, are diffusely branched and often have prostrate or spreading stems; the flowers are clustered in the upper leaf axils. In the first species the stems have ascending hairs and the leaflets are broadly obovate; in the second the hairs are retrorse and the leaflets oblong.

L. violacea, Violet Bush Clover, has weakly ascending stems and long-stalked leaves with elliptic leaflets up to 1 $^1/_2$ inches long. The flowers are in a much-branched, loose inflorescence, the racemes longer than the leaves. The keel is longer than the wing petals.

L. virginica, Slender Bush Clover, is an erect plant with a few stiff, sometimes arching branches. Its leaves are crowded, with long petioles and leaflets that are linear-oblong, less than $^1/_4$ inch wide, and about 4 times as long. The flowers are in short clusters in the leaf axils.

L. intermedia, Wandlike Bush Clover (plate 249), is similar to the preceding species, but its long-stalked leaves have elliptic segments $^1/_4$ to $^3/_4$ inch wide. The stem is essentially smooth. Its flowers are in short axillary clusters, much more numerous and crowded at the top. **L. nuttallii,** which has long spreading hairs on the upper stem, is thought to be a hybrid between *L. intermedia* and *L. hirta.*

L. stuevei, Stueve's Bush Clover, resembles *L. intermedia* but its stem and leaves are densely covered with short, soft, spreading hairs. The leaflets are ovate-oblong and the flowers are in dense racemes in the upper axils.

L. capitata, Round-headed Bush Clover (plate 250). The leaves of this species are short-stalked and have elliptic segments 1 to 1 $^1/_2$ inches long (more than twice as long as wide). The creamy white flowers are marked with purple and are aggregated into dense, headlike globose clusters crowded into the upper axils.

L. hirta, Hairy Bush Clover, is a very hairy plant with distinct petioles and elliptic to obovate leaflets up to 1 1/2 inches long and more than half as wide. The flowers are yellowish white, in short cylindrical spikes raised on long peduncles well above the leaves.

L. cuneata, Sericea (plate 251), has long, leafy, wandlike branches. Its leaflets are narrowly wedge-shaped, ¹/₂ to 1 inch long. The flowers, white with purple markings, are solitary or in groups of 2 or 3 in the upper axils, shorter than the subtending leaves.

Sometimes called "Chinese Lespedeza," Sericea was introduced into the United States from Asia and placed under cultivation but escaped and now has become naturalized throughout the Southeast and elsewhere. This is just one more example of an imported species which, contrary to expectations, proved to be difficult to control and has become a threat to native vegetation.

Trifolium spp., Clover. Only three genera in the Bean Family—*Trifolium*, *Medicago*, and *Melilotus*—have leaves with toothed margins. Except for two species, all of our true Clovers have palmately compound leaves. They commonly grow in fields and waste places (sometimes invading lawns) and flower from spring until fall.

Flowers yellow: *T. aureum, T. campestre, T. dubium*
Flowers white, pink, or red: *T. repens, T. hybridum,*
  T. virginicum, T. carolinianum, T. reflexum, T. pratense,
  T. incarnatum, T. arvense

Trifolium aureum (*T. agrarium*), Hop Clover (plate 252), is an erect plant and our largest yellow Clover. Its leaflets are all virtually sessile, and the heads are short-cylindrical, ¹/₂ to ³/₄ inch long, on stalks from the upper axils. As the flowers age they become striate, brownish, and reflexed.

T. campestre (*T. procumbens*), Low Hop Clover, is a smaller, prostrate species. The terminal leaflet is distinctly stalked. There usually are 20 or more flowers in each head, which is ¹/₂ inch long.

T. dubium, Least Hop Clover. Also prostrate, this too has pinnate leaves but is smaller than the preceding species. There are between 3 and 15 flowers to a head.

T. repens, White Clover (plate 253), is a smooth plant with creeping stems that send up long-stalked leaves and separate leafless, flowering stalks. The heads are globose, ³/₄ to 1 inch in diameter, with white flowers sometimes tinged with pink. The leaflets are more or less elliptic, ³/₄ inch long, with a pale **V**-shaped mark near the middle.

T. hybridum, Alsike Clover (plate 254). Another smooth species, this is frequently confused with T. *repens*; however, it is not stoloniferous but has erect, leafy stems with heads of white to pink flowers arising on long stalks from the upper axils. The flowers turn brown with age.

T. virginicum, Kates Mountain Clover (plate 255), is one of the most distinctive and quite possibly the best known of the endemic plant species largely restricted to shale barrens in the northern part of our region. Its pubescent stems are prostrate but non-stoloniferous. The leaflets are highly unusual—narrowly oblong, up to 2 ¹/₂ inches long, but less than one-third as wide. Its flowers are dull white, the individual blossoms about ¹/₂ inch long.

T. carolinianum may be either erect or trailing. Its leaves are obovate and small, the leaflets only ⁵/₈ inch long. The flowers are purplish, in ¹/₂-inch globose heads on long, slender stalks from the axils.

T. reflexum, Buffalo Clover. Larger than most of the preceding species, this plant may have leaflets more than 1 ¹/₂ inch long and globose heads about 1 inch across. The flowers are white or pink with a deep rose standard.

T. pratense, Red Clover (plate 256). Probably our best-known Clover, its flowers are actually rose-purple (occasionally white); the heads are large and sessile or very short-stalked. Its leaflets are widest near the middle and have a lighter green **V**-shaped blaze.

Of all the plants that have come to America from Europe, Red Clover must be counted among the most valuable, giving the phrase "in clover" a basis in fact. It is planted extensively as pasture and for hay, and as a "green manure" to be plowed under to enrich the soil.

T. incarnatum, Crimson or Italian Clover (plate 257). An imported forage plant that often escapes, this has large, dense heads of crimson flowers, ovoid at first, ultimately cylindrical, and up to 3 inches long. Its leaflets are broadly obovate.

T. arvense, Rabbit Foot Clover (plate 258). The inflorescence of this plant appears grayish and furry due to the long, soft hairs of the calyxes, which all but conceal the tiny pink or white flowers. The stalked heads are numerous and short-cylindrical. The leaflets are very narrowly oblanceolate, not more than ³/₄ inch long, minutely toothed only near the apex.

Medicago lupulina, Black Medick (plate 259). Our species of *Medicago* differ from *Trifolium* in having flowers that fall off soon after blooming, revealing curved legumes, instead of persistent flowers that conceal the short, straight fruits of the Clovers. Black Medick has small heads of yellow flowers and pinnately trifoliolate leaves. The fruits are blackish and kidney-shaped. Lawns and waste places. Spring–summer.

M. sativa, Alfalfa, Lucerne. This well-known agricultural plant often escapes to roadsides and waste areas. Its flowers are blue-violet, in short racemes, and the leaves are pinnate. The legumes are circular and loosely coiled. Spring–summer.

Melilotus alba, White Sweet Clover. The Sweet Clovers are tall plants with fragrant foliage and long-stalked, slender racemes of flowers less than ¹/₄ inch long. The leaflets are narrow, stalked, and finely toothed. Except for the color of its flowers, **M. officinalis,** Yellow Sweet Clover (plate 260), is virtually identical. Roadsides and waste places. Summer–fall.

Leaflets more than 3

Apios americana, Groundnut (plate 261), is a high-climbing, twining vine with pinnate leaves; the 5 or 7 leaflets are ovate, tapering to a sharp point, about 2 inches long. The flowers are ¹/₂ inch long and fragrant and in dense, compact racemes in the axils; the petals are thick and brownish purple. Moist, wooded bottom-lands. Summer.

The name "Groundnut" refers to the small, edible underground tubers. They were mentioned as food in a 1590 account by Thomas Hariot, who was a member of an expedition sent by Sir Walter Raleigh to Roanoke Island in what is now North Carolina, and would become the author of the first English book to be written in America.

Lotus corniculatus, Bird's-foot Trefoil (plate 262), is a low-growing plant with bright yellow $^1/_2$-inch flowers in long-stalked, headlike umbels. The pinnate leaves have 5 leaflets, the lowermost pair attached at the base of the petioles and resembling stipules. Fields and roadsides. Summer.

Coronilla varia, Crown Vetch (plate 263), has been planted extensively on road banks to control soil erosion and frequently escapes to fields. Its flowers are $^1/_2$ inch long, pink varying to white, in long-stalked umbels. The leaves are odd-pinnate with from 9 to many more oblong leaflets $^1/_2$ to $^3/_4$ inch long. Spring–summer.

Tephrosia virginiana, Goat's Rue (plate 264), is an erect, unbranched plant covered with silky white hairs. Its leaves are odd-pinnate with 15 to 25 oblong leaflets 1 inch or longer. The flowers are $^1/_2$ to $^3/_4$ inch long, the nearly circular standard creamy white, the connivent wings and keel rose-purple; they are borne in a single, crowded terminal raceme. Dry, open woods and roadsides. Late spring–summer.

T. spicata is decumbent, branched, and has few-flowered racemes at the ends of long, slender axillary or terminal stalks. They are white at first, soon turning pink, then reddish purple. Open woods, roadsides, and clearings. Spring–summer.

Astragalus canadensis, Milk Vetch (plate 265). This enormous genus is represented here by a single species. It is an erect plant with odd-pinnate leaves; the 15 to 31 leaflets are oblong and up to 2 inches long. The flowers are yellowish white, $^1/_2$ inch long, and slender (the standard is narrow, in contrast with *Tephrosia*). They are in thick, densely flowered axillary spikes. Moist woods and thickets. Summer.

Lupinus perennis, Wild Lupine (plate 266). This handsome wildflower, a close relative of the famous Texas "bluebonnets," is our only member of the Bean Family with leaves palmately divided into more than 3 segments—actually there are from 7 to 11. It may grow to 2 feet high, and has ¹/₂-inch blue flowers in erect racemes. Open woods and clearings. Spring–early summer.

Lupinus subcarnosus *was selected as the state flower of Texas in 1901, but there ensued a long period of uncertainty and controversy about the choice. The matter was finally put to rest in 1971, when the legislature officially broadened the designation to include all species native to the state, which in-clude* L. texensis *and* L. havardii.

Oxalidaceae, Wood Sorrel Family

Oxalis spp., Wood Sorrel. Members of this genus are recognizable by their leaves, which are palmately divided into 3 obcordate leaflets. The flowers have 5 separate petals, often notched.

Oxalis montana *(O. acetosella),* Common Wood Sorrel (plate 267), has ¹/₂- to ³/₄-inch white flowers veined in deep pink. They are borne singly on short, naked stalks rising only slightly above the leaves, which frequently are massed. Rich woods, usually at high elevations. Spring.

O. violacea, Violet Wood Sorrel (plate 268), grows up to 8 inches high, and the flowers are on leafless stems but in umbelliform clusters. They are rose-purple and about ¹/₂ inch across; each sepal has an orange gland at its apex. Woods. Spring–early summer.

O. grandis, Large Wood Sorrel (plate 269). The size of this species makes it easy to identify. The yellow flowers are ³/₄ to 1 inch across and the leaflets, which are edged in reddish purple, may be as much as 2 inches wide. Woodlands. Spring–summer.

O. stricta, Yellow Wood Sorrel. There is much confusion surrounding the taxonomy of our other species, which are extremely variable. *O. stricta* is usually described as having cymose inflorescences and pedicels that are abruptly deflexed in fruit, and flowering from late spring to fall in waste areas and lawns, and **O.** *europaea* as being similar but with erect fruiting stalks. Both have ⁵/₈-inch yellow flowers on leafy stems.

O. *dillenii* (*O. florida*) differs in having its flowers in umbels, and begins flowering in early spring.

GERANIACEAE, GERANIUM FAMILY

Geranium maculatum, Wild Geranium, Cranesbill (plate 270). Our species of *Geranium* have mostly palmately lobed or dissected leaves (the segments themselves toothed or lobed) and 5-parted pink to purple flowers with 10 stamens. The largest is the spring-flowering G. *maculatum*, which grows to 1 foot or more in height and has several rose-purple flowers 1 to 1 ¹/₂ inches wide in a loose cyme and leaves up to 3 inches wide that are deeply 3- to 5-lobed. Rich woods.

G. *columbinum* is a somewhat smaller, sprawling, alien herb. Its flowers are in pairs; both the peduncles and the pedicels are elongate. The petals are ³/₈ inch long and purplish pink. The leaves, up to 2 inches wide, are 5- to 7-cleft into narrowly lobed divisions. Late spring–summer.

Another introduced species is **G.** *sibiricum,* Siberian Cranesbill (plate 271). Its palmate leaves are mostly 3-cleft, the segments coarsely toothed or lobed toward their summits. The calyx lobes terminate in a conspicuous awn. The ¹/₄-inch petals are pale pink with darker veins, not notched.

The erect, freely branched Carolina Cranesbill, **G.** *carolinianum* (plate 272), is a common native weed of dry, sandy, disturbed sites and flowers during spring and summer. It has pale pink to nearly white flowers with petals about ³/₈ inch long and notched at the apex, crowded into compact clusters on short stalks. Its leaves are deeply cleft, the segments slender. **G.** *dissectum* is similar but has deep pink or purple flowers with ¹/₄-inch petals.

G. robertianum, Herb Robert (plate 273), differs markedly from the others in having its leaves divided into 3 distinct segments of which the terminal one is stalked. Its flowers are rose-purple, $^1/_2$ inch wide. It favors damp, rocky woods and blooms from late spring until fall.

Two sprawling introduced species are summer-flowering lawn weeds. Both have round leaves under 2 inches across and numerous flowers with $^1/_4$-inch, notched rose-pink petals. In *G. molle* the leaves have 5 to 9 short, blunt lobes, while in *G. pusillum* they are more deeply cleft into wedge-shaped segments. The latter is unusual in having only 5 stamens.

Erodium cicutarium, Storksbill, Filaree (plate 274). *Erodium* differs from *Geranium* in that the leaves are elongate and are pinnately compound, a combination of features that gives them a fernlike appearance. In our species the flowers have magenta-pink oval petals without a notch at the apex. Fields and waste places. Spring–summer.

LINACEAE, FLAX FAMILY

Linum spp., Wild Flax. Our species of *Linum* are slender plants, branched above, with narrow leaves and 5-petaled yellow flowers about 1/4 inch across, and 5 stamens, blooming in summer. (They may be separated from Frostweed, *Helianthemum*, by the larger petals, numerous stamens, and presence of cleistogamous flowers in the latter.)

Linum striatum (plate 275) has mostly opposite leaves, although some of the upper ones may be alternate. The inflorescence is paniculately branched.

L. virginianum has a corymbose inflorescence, and most of its leaves are alternate. Other species with narrower leaves (some of them regarded at one time as varieties of *L. virginianum*, and all difficult to distinguish) may be seen less frequently in the mountains.

POLYGALACEAE, MILKWORT FAMILY

Polygala paucifolia, Gay Wings, Fringed Polygala, Flowering Wintergreen (plate 276). Since they are much larger than those of the other species, the individual flowers of *P. paucifolia* provide an excellent means of examining the peculiar floral structure that is common to all the Milkworts. The conspicuous parts are 1) the corolla tube, which consists of 3 united petals, the lower one fringed at its summit, and 2) the wings, which are a pair of similarly colored spreading sepals. *P. paucifolia* is a low plant bearing 1 to 4 of these flowers. They are rose-purple, each wing and the corolla ½ to ¾ inch long. The oval leaves are clustered near the top of the stem.

The remaining species have small but more numerous flowers. Most inhabit open woods and fields.

> Leaves whorled: *P. verticillata, P. cruciata*
> Leaves alternate; stems several: *P. senega, P. polygama*
> Leaves alternate; stems solitary: *P. incarnata, P. sanguinea,*
>   *P. curtissii, P. nuttallii*

P. verticillata, Whorled Milkwort, has linear leaves in whorls of 3 to 6 on the lower part of the stem, which is divergently branched above. The white or pinkish flowers are minute, in long-stalked, tapering racemes less than ½ inch long. Summer–fall.

P. cruciata, Cross-leaved Milkwort, Drum Heads. In this species the leaves are mostly in whorls of 4. The flowers are pinkish, in dense, cylindrical, nearly sessile headlike racemes ½ inch thick. The wings are broadly triangular. Wet habitats. Summer–fall.

P. senega, Seneca Snakeroot (plate 277). The lanceolate cauline leaves of Seneca Snakeroot are unusually broad for a small-flowered *Polygala*. Several unbranched stems arise from the base and terminate in long, narrow, uninterrupted racemes of white or greenish flowers. Spring–early summer.

P. polygama, Racemed Milkwort, has several decumbent stems with numerous narrow leaves about 1 inch long. The flowers are rose-purple, and are in a loose, slender raceme. Spring–summer.

P. incarnata, Pink Milkwort. The long, conspicuously fringed corolla tube of *P. incarnata* is distinctive: At ¼ to ³/₈ inch, it is more than twice the length of the wings. The flowers are pink, in dense spikes. There are relatively few linear leaves scattered along the stem. Summer.

P. sanguinea, Field Milkwort. In contrast to the preceding species, the wings of Field Milkwort are longer than the corolla. Its flowers are usually rose-purple, and are aggregated into dense, oval, headlike racemes ³/₈ inch thick. In this and the two following, the axis is made rough by the bracts, which persist after the flowers drop off. Summer–fall.

P. curtissii, Curtiss' Milkwort (plate 278). Here the wings are about the same size (¹/₈ to ³/₁₆ inch) as in the above but are about equaled by the corolla. The flowers are rose-purple with a yellow tip in dense, oblong heads. Nuttall's Milkwort, **P. nuttallii,** is similar but smaller, the wings under ¹/₈ inch and the racemes only ³/₁₆ inch thick. Summer–fall.

EUPHORBIACEAE, SPURGE FAMILY

Euphorbia **spp.,** Spurge. In *Euphorbia,* the small structure that appears to be a single flower—technically known as a cyathium—is a lobed, cup-shaped involucre with glands at the bases of the segments, containing one or more staminate flowers (each consisting of a single stamen only) surrounding a solitary pistillate flower (a stalked ovary topped by 3 styles). In some species the involucre also bears appendages that simulate petals. (The well-known Poinsettia is a *Euphorbia* in which brightly colored leaves beneath the inflorescence give the impression of a large-petaled flower.) All species of *Euphorbia* contain a milky sap. In addition to those described below, we have a number of species that have minute cyathia and therefore fall somewhat short of satisfying the popular conception of a wildflower.

Euphorbia corollata (*E. zinniiflora*), Flowering Spurge (plate 279), takes its name from, and is made attractive by, the 5 bright white, rounded, petal-like appendages on the rim of the involucre; the entire structure is about ¹/₄ inch across. Its stem leaves are elliptic, 1 to 2 inches long, in whorls subtending the branches, alternate below. Open areas. Spring–fall.

E. purpurea, Glade Spurge (plate 280), is a stout 3-foot plant with alternate blunt, oblong stem leaves up to 4 inches long and 1 inch broad. There are also small kidney-shaped bracteal leaves in the inflorescence. The cyathium has 5 segments which may be tinged with purple (accounting for the specific epithet). Woods. Spring–summer.

Caper Spurge, **E. lathyris,** has opposite, pointed, lanceolate stem leaves and similar but shorter bracteal leaves. Waste places. Summer.

E. cyparissias, Cypress Spurge (plate 281), is frequently found in dense masses in waste areas and especially in old cemeteries. The light green leaves are linear, seldom more than 1 inch long, very numerous, and crowded. There is a many-rayed umbel of "flowers" with broadly ovate bracteal leaves on the pedicels, all greenish yellow. Spring.

E. commutata, Wood Spurge, is smooth and openly branched, with $^1/_2$-inch oblanceolate stem leaves, mostly alternate, and pairs of broadly ovate leaves in the inflorescence. The small, pale yellow glands are crescent-shaped with a short spur projecting from each end. Usually in basic soils. Spring–summer.

The interesting flower structure of the *Euphorbias* can be observed on a reduced scale in several weedy species, of which **E. maculata** (*E. supina*), Wartweed, is our most common representative. Although sometimes erect, it is most distinctive in its prostrate form in which many hairy stems radiate from a central point, where often it will be rooted in the soil beneath pavement cracks. Its dark green, purple-blotched leaves are opposite, toothed, and oblong but oblique at the base. The cyathia are axillary, with 4 small white petaloid appendages. Spring–fall.

Buxaceae, Box Family

Pachysandra procumbens, Allegheny Spurge (plate 282). The stems of Allegheny Spurge arise from long rhizomes, sometimes forming extensive patches. The 3-inch ovate leaves are long-petioled, bright green at first but becoming darker and mottled with age, have coarse teeth in the upper portion, and are crowded near the summit. One or more erect spikes of flowers, 2 to 4 inches long, grow from the lower part of the stem. Each spike bears staminate flowers above, made conspicuous by their thick white filaments, and pistillate flowers, if any, below. Rich woods, usually calcareous. Early spring.

P. terminalis, Japanese Spurge, is commonly planted as an ornamental ground cover. Its spikes are terminal, and the leaves are uniformly green.

CELASTRACEAE, STAFFTREE FAMILY

Euonymus americanus, Strawberry Bush (plate 283). Our native species of *Euonymus* are semi-evergreen shrubs with opposite leaves, noted more for their colorful autumn fruits than for the flowers that appear in the spring. It is the former that gave *E. americanus* the affectionate name of "Hearts a-bustin' with Love." This is an erect plant with bluish green stems and serrate, lanceolate to ovate leaves. The $^1/_2$-inch flowers have 5 yellowish or brownish green ovate petals; they grow singly or in twos or threes on long stalks from the leaf axils. The fruit is a warty, bright crimson capsule $^1/_2$ to $^3/_4$ inch in diameter. At maturity this splits into 5 segments, which then spread apart to disclose the orange arils.

Trailing Strawberry Bush, **E. obovatus** (plate 284), is similar but prostrate and has leaves that are widest above the middle. Its fruits are usually 3-lobed. Both species will be found in rich woods.

Burning Bush, **E. atropurpureus,** is a tall shrub with 4-petaled, brownish-purple flowers, $^1/_4$ inch wide, in cymes. The fruit is smooth and the aril red. Woodland margins. Late spring.

BALSAMINACEAE, TOUCH-ME-NOT FAMILY

Impatiens spp., Touch-me-not. The irregular, cornucopia-like flowers of our *Impatiens* dangle from threadlike stalks and consist of 3 sepals, the upper 2 small and the lower forming a broad, funnel-shaped sac and extending backward into a slender spur. There is 1 upper petal and a pair on each side joined and appearing to be single two-lobed petals. When mature, the fleshy seed capsules burst elastically with startling suddenness at the slightest touch. The plants are succulent and grow to a height of several feet.

Impatiens capensis (*I. biflora*), Spotted Jewel Weed (plate 285), has orange flowers copiously spotted with reddish brown. In moist places. Summer–fall.

Pale Jewel Weed, **I. pallida** (plate 286), is very similar but has lemon yellow (sometimes cream-colored or white) flowers that may be sparingly spotted. Usually at higher elevations. Summer–fall.

Rubbing the copious juice from Jewel Weed stems onto the skin is reputed to afford protection from Poison Ivy and to relieve the discomfort resulting from contact with it or with stinging Nettles. It is also used to treat some forms of fungal dermatitis.

MALVACEAE, MALLOW FAMILY

Hibiscus moscheutos, Rose Mallow. This interesting family includes the plants which are the source of cotton, okra, and—until it was synthesized—marshmallow. The column formed by the stamens united around the style is a distinctive feature and is especially conspicuous in the large flowers of this genus. *H. moscheutos* may grow as tall as 8 feet and has lanceolate leaves (some may be obscurely 3-lobed), downy beneath; usually there is also a leaf on the peduncle. The flowers are 6 inches across, white or cream-colored (rarely pink), and red at the base. Plants with ovate leaves, leafless peduncles, and flowers that are usually pink are considered to be **ssp. palustris,** Swamp Rose Mallow (plate 287). Wet places. Summer–fall.

About as tall is **H. laevis** (*H. militaris*), Halberd-leaved Marsh Mallow, in which the leaves are smooth beneath and spear-shaped, with 2 sharp, divergent basal lobes. Its flowers are also large, pink with a darker center. Stream banks. Summer.

H. trionum, Flower-of-an-Hour, is a low plant with leaves that are deeply parted into 3 narrow, coarsely toothed segments. Its flowers are 1 $^{1}/_{4}$ inches wide, pale yellow with a purple "eye," and remain open only for an hour or two. Fields and waste areas. Summer–fall.

Malva moschata, Musk Mallow (plate 288). This is an attractive plant with pink or white flowers and leaves that are deeply divided, the segments themselves deeply pinnatifid. The petals are about 1 inch long, triangular, and (as in all members of this genus) indented on the outer margin. Fields and roadsides. Late spring–summer.

M. sylvestris, High Mallow, may be 3 feet tall. Its leaves are roundish and shallowly lobed. The flowers are 2 inches across, pink with red veins. Disturbed sites. Late spring–early summer.

M. neglecta, Common Mallow, Cheeses (plate 289). This is a common weed around dwellings and in waste areas. It is a creeping plant with very long-stalked, nearly circular, shallowly lobed leaves. The $^{3}/_{4}$-inch flowers are white, pink, or lavender, on long stalks in the leaf axils. Spring–fall.

Sida spinosa, Prickly Mallow (plate 290), is a freely branched plant, up to 2 feet high, with lanceolate leaves and $^{1}/_{2}$-inch pale yellow flowers on peduncles about as long as the short leafstalks. The specific epithet refers to the short spine at the base of each petiole. **S. rhombifolia** can be distinguished by its wider rhombic leaves, pedicels many times longer than the petioles, and the absence of spiny stipules. Disturbed sites. Summer–fall.

Modiola caroliniana is a sprawling, hairy plant with 3- or 5-lobed, irregularly toothed leaves. The $^{3}/_{8}$-inch flowers, which open only in bright sunlight, are solitary on long stalks from the leaf axils; the petals are orange-red with a dark purple base. Waste places. Spring–summer.

Abutilon theophrasti, Velvet Leaf, Indian Mallow, is a stout, tall (to 5-foot) plant, soft-hairy throughout. Its leaves are heart-shaped with a tapering point, 4 to 6 inches long. The flowers are less than 1 inch wide, with bright yellow petals, growing in the leaf axils. Fields and waste areas. Summer–fall.

HYPERICACEAE, ST. JOHN'S-WORT FAMILY

Hypericum spp., St. John's-wort. Our species of *Hypericum* may be divided into two groups: 1) those whose flowers have 4 petals and either 2 sepals or 4 unequal ones (formerly in the genus *Ascyrum*); and 2) those with 5 petals and 5 sepals.

> Flowers 4-petaled: *H. crux-andreae*, *H. hypericoides*,
> *H. stragalum*
> Flowers 5-petaled: *H. graveolens*, *H. mitchellianum*,
> *H. perforatum*, *H. punctatum*, *H. denticulatum*, *H. canadense*,
> *H. mutilum*, *H. buckleyi*, *H. gentianoides*, *H. drummondii*

Hypericum crux-andreae (*H. stans*, *Ascyrum stans*), St. Peter's-wort (plate 291), is a small, shrubby plant with opposite elliptic leaves about 1 inch long, rounded, or clasping at the base. Its flowers are pale yellow with 4 obovate petals $^1/_2$ to $^3/_4$ inch long, and 3 or 4 styles. There are 2 sets of sepals, the outer pair quite large and the inner pair shorter and very narrow. Dry woods. Summer–fall.

H. hypericoides (*Ascyrum hypericoides*), St. Andrew's Cross (plate 292) resembles the above but has leaves that are narrower, less than 1 inch long, and tapered at the base. The flowers have 4 narrow oblong petals 5/8 inch long that form an oblique cross. There are only 2 styles. The sepals are very unequal, the inner pair being minute or even lacking. A similar plant with decumbent stems forming low mats is **H. stragalum** (*Ascyrum hypericoides* var. *multicaule*). Dry woods and open areas. Summer.

---

*The preceding two species are unusual in having their petals so oriented as to form a flattened "**X**," which is the shape of the cross of St. Andrew (as well as that of St. Patrick), in contrast to the right-angled form which is exemplified by St. George's cross. These are well illustrated by the Union Jack of Great Britain, in which the three crosses are superimposed to represent England, Scotland, and Ireland.*

H. graveolens, Mountain St. John's-wort (plate 293). Our 5-petaled species have opposite, entire, and pellucid- or black-dotted leaves and yellow flowers. *H. graveolens* is a handsome species, with elliptic leaves up to 2 ¹/₂ inches long and flowers more than 1 inch wide. There are numerous long stamens in several fascicles. Similar but smaller in all respects is **H. mitchellianum,** which has more densely flowered cymes and black lines or spots on the sepals and occasionally on the petals. Both species are found at high elevations, and natural hybridization is likely. Summer.

H. perforatum, Common St. John's-wort (known in the western United States as Klamath Weed) (plate 294). This is a much-branched plant and has short (1-inch), narrow, oblong leaves with translucent dots. (A photosensitzing pigment in these tiny glands was responsible for extensive livestock poisoning in California during the first half of this century.) The flowers are 3/4 to 1 inch wide, with black "stitching" on the margins of the petals. Roadsides and fields. Summer–fall.

Spotted St. John's-wort, **H. punctatum,** is sparingly branched and the leaves are longer (up to 2 inches) and wider, marked with black dots. The petals and sepals are copiously black-spotted. Thickets. Summer–fall.

As its common name indicates, Coppery St. John's-wort, **H. denticulatum,** has coppery yellow flowers; they are less than ³/₄ inch across and have numerous stamens. This species has a 4-angled stem and 1-inch linear ascending leaves, and is widely branched above. Rock outcrops and sandy soils. Summer.

H. canadense is a slender, well-branched plant with spreading linear leaves 1 inch long but only ¹/₁₆ inch wide. Its flowers are ¹/₄ inch across. Moist habitats. Summer–fall.

Dwarf St. John's-wort, **H. mutilum,** is another small-flowered species. It is diffusely branched and bears numerous flowers. The leaves are ovate to elliptic, up to 1 ¹/₂ inches long and ¹/₂ inch wide and rounded at the base. The flowers are less than ¹/₄ inch across. Wet soil. Summer–fall.

H. buckleyi (plate 295), Appalachian St. John's-wort, is an unusual southern Appalachian endemic. It is a decumbent shrub forming extensive mats on rocky cliffs, seepage slopes, and balds. The leaves are elliptical to obovate, and the flowers, solitary or few at the ends of the branches, are about ³/₄ inch wide. Summer.

The next two species have flowers so minute that some field guides ignore them. In **H. gentianoides,** known as Pineweed or Orange Grass (plate 296), they are ¹/₈ inch wide and are borne nearly sessile at the nodes. This plant is repeatedly branched into many extremely slender stems; the leaves are scalelike, appressed, and even smaller than the flowers. Similar to this is **H. drummondii,** which has fewer, more erect branches and ascending linear leaves about ¹/₂ inch long. The tiny flowers are on short stalks in the axils. Both this and the preceding species may be found on rock outcrops and roadsides and in dry fields. Summer–fall.

Triadenum virginicum (*Hypericum virginicum*), Marsh St. John's-wort (plate 297). The ¹/₂- to ³/₄-inch flowers of *Triadenum* are strikingly different from those of the *Hypericums* in that they are pink and have 9 stamens divided equally among 3 fascicles, which alternate with orange glands. In this species the leaves are oval with cordate-clasping bases, 1 to 2 inches long and at least one-third as wide. The sepals are pointed. Bogs, marshes, and wet meadows. Summer.

T. walteri (*T. tubulosum* var. *walteri, Hypericum walteri*) has similar but smaller flowers, less than ¹/₂ inch across, with blunt sepals. Its leaves are larger (to 4 inches) and narrowed at the base to short petioles. Low, wet woods and marshes. Summer–early fall.

Cistaceae, Rockrose Family

Helianthemum spp., Frostweed. The Frostweeds are plants with small, entire, elliptic, alternate leaves. Most species bear several chasmogamous flowers ³/₄ to 1 inch across, with 5 yellow petals and numerous stamens (which tend to lie to one side) as well as many smaller cleistogamous (apetalous) flowers in the upper axils. The chasmogamous flowers open only in bright sunshine for a few hours, usualy in the morning. Frostweeds prefer dry, sandy habitats and bloom in spring or summer.

In **Helianthemum bicknelli** the broad outer sepals of the chasmogamous flowers are about equal in length to the narrow inner ones, while in **H. propinquum** they are much shorter. **H. canadense** (plate 298) has only 1 or 2 such flowers, and they may be up to 1 ¹/₂ inches wide.

Lechea racemulosa, Pinweed (plate 299). These diffusely branched plants bear a multitude of tiny flowers with 3 red petals which expand only briefly at anthesis but persist after withering. The 3 white plumose stigmas are conspicuous despite their small size. This is our most widespread species. Dry, rocky soil. Summer.

VIOLACEAE, VIOLET FAMILY

Viola spp., Violet. There are many more kinds of Violets than most of us suspect, and identifying them (especially the blue-flowered ones) is complicated by the tendency of closely related species to hybridize and by uncertainty on the part of even the experts as to where to place some of these intermediate taxa.

Violets have bilaterally symmetric flowers with 5 slightly unequal petals, the lower one extended backward into a spur. They bloom in spring, with cleistogamous (apetalous) flowers ensuing throughout the summer months. It is customary to divide the Violets into two broad categories: 1) the "stemmed" species—i.e., those with leafy stems and bearing flowers in the axils—and 2) the "stemless" species, in which the leaves and the flowers are on separate stalks arising directly from rhizomes or stolons.

Plants leafy-stemmed; flowers blue: *V. rostrata, V. walteri,*
  V. conspersa, V. rafinesquii
Plants leafy-stemmed; flowers white: *V. rafinesquii,*
  V. canadensis, V. striata
Plants leafy-stemmed; flowers yellow: *V. pubescens,*
  V. tripartita, V. hastata, V. arvensis
Plants stemless; flowers blue to purple: *V. pedata, V. palmata,*
  V. triloba, V. septemloba, V. sagittata, V. fimbriatula, V. sororia,
  V. cucullata, V. affinis, V. septentrionalis, V. hirsutula
Plants stemless; flowers white: *V. blanda, V. pallens,*
  V. primulifolia, V. lanceolata
Plants stemless; flowers yellow: *V. rotundifolia*

Plants leafy-stemmed; flowers blue

Viola rostrata, Long-spurred Violet (plate 300). An infallible field mark of this species is its spur, which is slender and measures $^1/_2$ inch or more in length. The flowers are light blue-violet with a dark eye, the lateral petals beardless. The leaves are heart-shaped and pointed at the apex. Rich woods.

A southern species, **V. walteri,** is a finely pubescent plant with several prostrate stems rooting at the nodes. Its leaves are roundish, blunt at the tip, purple beneath. The violet-colored flowers have a spur less than $^1/_4$ inch long, and the lateral petals are bearded. Dry woods.

Dog Violet, **V. conspersa,** shares a number of characters with the preceding, but it is smooth and has erect stems that do not root at the nodes. The leaves are green beneath. Low woods and meadows.

Plants leafy-stemmed; flowers white

V. rafinesquii *(V. kitaibeliana)*, Wild Pansy (plate 301). A slender, delicate plant, Wild Pansy has small, long-stalked white (or blue) flowers with short, blunt spurs and short sepals. The leaves also are small and have round to spatulate blades with scalloped margins; the stipules are leaflike and deeply lobed. Fields, lawns, and roadsides.

V. canadensis, Canada Violet (plate 302). Our tallest Violet, this species has white flowers (violet-tinged beneath) with a yellow eye and purple veins near the base of the petals. The flower stalks are shorter than the leaves, which are heart-shaped with long-pointed tips. The stipules are small and entire. Cool woodlands.

V. striata, Pale or Cream Violet. This has ivory-white flowers (not tinted on the back) with brownish purple veins, on stalks that are longer than the leaves. Its stipules are large and have a fringe of sharp teeth, and the sepals are ciliate. Alluvial woods and meadows.

Plants leafy-stemmed; flowers yellow

V. pubescens, Downy Yellow Violet (plate 303), is soft-hairy throughout, and has no basal leaves (or occasionally a solitary one). **Var. *leiocarpon*** (*V. pensylvanica*), Smooth Yellow Violet, is essentially hairless and usually has several long-stalked basal leaves. Together, these constitute our only leafy-stemmed yellow Violets with heart-shaped leaves. Woodlands.

V. *tripartita* (plate 304) is unusual in having at least some of its leaves deeply dissected or lobed into 3 segments. In **var. *glaberrima*** the leaves are not divided, and have truncate or wedge-shaped—but never cordate—bases. In these and the next species the yellow petals are purplish on the back. Dry woods.

V. *hastata*, Halberd-leaved Violet (plate 305). This species bears only 2 to 4 leaves, all high on the stem. They are triangular, about twice as long as wide, have halberd-shaped bases, and often are silvery-mottled between the veins. Woodlands.

V. *arvensis*, Field Pansy, is quite similar to *V. rafinesquii* but has pale yellow flowers sometimes marked with purple, and a slender spur. Fields and waste places.

---

*The familiar "Johnny-jump-up" or "Heartsease" (*V. tricolor*) occasionally escapes from gardens but seldom persists as a member of our flora. It resembles* V. arvensis *but has short sepals and is more colorful, as befits the principal ancestor of the cultivated pansy.*

Plants stemless; flowers blue to purple

V. *pedata*, Birdfoot Violet, has the largest flowers of all—from $^3/_4$ to 1 $^1/_2$ inches across. In the typical species they are bicolored, the 2 upper petals deep violet and the others lavender, but in the more common **var. *lineariloba*** (plate 306) they are a uniform pale lavender. This species is unique in that the orange tips of the stamens protrude conspicuously from the throat. Its leaves are palmately divided into 3 main segments, the lateral ones again deeply parted into narrow lobes often bearing teeth at the apex. Dry, rocky woods, fields, and road banks.

V. *palmata*, Early Blue or Wood Violet. In this and the next two species at least some of the leaves are deeply but more or less irregularly lobed or dissected (although unlobed cordate leaves may be present at any time). In the case of *V. palmata* there are from 5 to 11 rather narrow segments, the middle one the widest. The flowers are normally deep reddish violet, but there is a variant with streaked violet-and-white petals known as **forma striata** (plate 307). Woodlands.

V. triloba (*V. palmata* var. *triloba*), Three-lobed Violet. A very similar species, this usually has leaf blades with 3 or sometimes 5 broad lobes, the lateral ones frequently bearing coarse teeth; the sinuses are occasionally shallow.

V. septemloba (*V. esculenta*) is primarily a coastal species, growing under pines. It has variously 3- to 5-lobed leaves, the median segment longer or wider. The leaf undersides and petioles tend to be less hairy than in the preceding two species.

V. sagittata (*V. emarginata*). The next two have leaf blades that are distinctly longer than wide and essentially entire except for teeth or relatively short lobes confined to the base. In *V. sagittata* the leaves are smooth, purplish beneath, and lanceolate with a sagittate base, the flaring lobes incised or deeply toothed. The petioles are longer than the leaf blades. Moist meadows.

V. fimbriatula, Downy Violet (plate 308). This is a small, densely hairy plant with petioles shorter than the leaf blades. The leaves are ovate at first, later becoming oblong-ovate with shallowly crenate margins and small rounded (but not flaring) basal lobes. Dry, open areas.

V. sororia, Common Blue Violet (plate 309), is at once the most widespread, universally admired, and easily recognized of all our Violets. Many of us can describe it from memory: smooth, heart-shaped leaves, long, sleek stems overtopping the inch-wide, blue-violet flowers with a white eye and bearded with tufts of hairs on the lateral petals. Some learned to know its scientific name, *Viola papilionacea*, but that has been replaced, and we are told that it is now to be called *Viola sororia* and, in fact, that all of our "stemless blues" might well belong to a single polymorphic species. Included in this would be the familiar Confederate Violet, **var. priceana,** with its grayish white flowers and an eye-spot of violet-colored veins.

V. cucullata, Marsh Blue Violet (plate 310), differs in having its flowers on long stalks much surpassing the leaves; they are light blue-violet with a darker eye-spot. The lateral petals are bearded, but the hairs are knobbed. Wet places.

Like the two preceding species, Le Conte's Violet, **V. affinis** (*V. floridana*), is smooth and has long petioles, but the leaf blades are more triangular, longer than wide, and taper to a long point. The flowers are violet with conspicuous white centers, and all 3 lower petals are villous at their bases. Low, moist woods.

V. septentrionalis, Northern Blue Violet. Native to the Northeast where it is often cultivated, this species occurs infrequently in the South, mostly at high elevations. It is pubescent and has extremely long petioles. All of the lower petals are bearded.

V. hirsutula, Southern Wood Violet, is a small plant with leaves under 2 inches wide that lie close to the ground. They are silvery-downy with purple veins above, smooth and purplish beneath, and blunt at the apex. The flowers are bright reddish purple, and the lateral petals have club-shaped hairs. Dry woods.

Plants stemless; flowers white

V. blanda (*V. incognita*), Sweet White Violet (plate 311). The most common white Violet in the southern mountains, its leaves are nearly circular, cordate at the base with a narrow sinus and short-pointed at the apex, the upper surface satiny. The flowers are $^1/_2$ inch long or more on reddish peduncles, the upper petals oblong, twisted, and bent backward. Moist woods, usually under conifers.

Less frequently encountered in our area is **V. pallens** (*V. macloskeyi* var. *pallens*). Its leaves have a blunt apex and a more open basal sinus and lack the satiny sheen of the preceding species. The flowers are less than $^1/_2$ inch long, the upper petals are obovate and not twisted, and the peduncles are green. Cool streamsides and other wet habitats.

V. primulifolia, Primrose-leaved Violet. The leaves serve to identify this species. The blades are oblong to ovate, pinnately veined, with a truncate or rounded base tapering to a winged petiole.

In contrast, Lance-leaved Violet, **V. lanceolata,** has narrow leaf blades 2 to 3 times as long as wide, tapering very gradually to a reddish petiole. Both species are found in open, moist areas.

Plants stemless; flowers yellow

V. rotundifolia, Round-leaved Yellow Violet (plate 312). Quite possibly the easiest of our Violets to identify, V. *rotundifolia* begins to bloom in early spring just as its leaves are beginning to unfold. They have crenate margins and continue to expand well into summer, becoming broadly ovate and up to 4 ½ inches across, and lie almost flat. The lower petals have brown veins. Rich woods.

Hybanthus concolor (*Cubelium concolor*), Green Violet (plate 313). Outwardly, the so-called Green Violet bears little resemblance to the Violets with which we are familiar and which are in the genus *Viola*. It is a coarse plant up to 3 feet tall with alternate elliptic leaves, in the axils of which small greenish flowers hang on short stalks. They are bilaterally symmetric, less than ¼ inch long, and have 5 petals and 5 sepals all about the same size except that the lower petal is slightly larger and has a saclike base. Usually on basic soils. Spring.

PASSIFLORACEAE, PASSION FLOWER FAMILY

Passiflora incarnata, Passion Flower, Maypop (plate 314). The intricate structure of a Passion Flower makes it impossible to mistake it for any other. In this species the 5 sepals and 5 petals are whitish or lavender and are surmounted by an elaborate, fringelike, purple-banded corona 2 inches or more in diameter. Above this, the 3 spreading styles are a conspicuous feature. This is a trailing or climbing tendril-bearing vine with deeply 3- to 5-lobed, finely serrate leaves; the flowers are borne singly in the axils. Dry fields and roadsides. Spring–summer.

Yellow Passion Flower, **P. lutea** (plate 315), climbs by tightly coiled tendrils. It has shallowly 3-lobed leaves without teeth and axillary flowers which are greenish yellow and only 1 inch across. Woods and thickets. Summer–fall.

CACTACEAE, CACTUS FAMILY

Opuntia humifusa (*O. compressa*), Prickly Pear
(plate 316). Cacti are hardly typical of the southern
mountain flora, but this one has a surprisingly wide
distribution in our region. It is not common, however, and it is the only species we can expect to en-
counter. *O. humifusa* is a sprawling plant with stems
composed of flattened, nearly circular, jointed pads.
These have regularly spaced tufts of tiny barbed hairs
(glochids), which detach at the slightest touch and
can be extremely painful; the long spines associated
with many cacti are absent, or occasionally occur singly. The attractive 2- to 3-inch yellow flowers are
made up of numerous sepals intergrading into petals.
Open sandy or rocky habitats. Summer.

LYTHRACEAE, LOOSESTRIFE FAMILY

Lythrum salicaria, Purple Loosestrife (plate 317).
In the Northeast this European introduction has
taken over large areas of marshland and margins of
rivers and lakes, crowding out many native plants,
but in the South it is relatively rare. It grows to 4
feet tall and has sessile leaves in pairs or whorls of 3.
The bright magenta flowers are radially symmetric,
$^1/_2$ to $^3/_4$ inch across, with 6 petals and 12 stamens,
and are densely crowded in tapered spikes. Summer–
fall.

L. alatum, Winged Loosestrife. This native species has
a 4-angled stem with opposite leaves below and alternate ones above. Its flowers differ from the preceding
species in being $^1/_2$ inch or less in width, having only 6
stamens, and being borne singly or paired in the axils.
Wet, open areas. Summer–fall.

Cuphea viscosissima (*C. petiolata*), Clammy Cuphea,
Blue Waxweed (plate 318). As the names indicate, this
is a viscid-hairy plant throughout. Its leaves are opposite, narrowly ovate, stalked, and up to 2 inches long.
The flowers are magenta, bilaterally symmetric, and usually solitary in the upper axils. The calyx tube is about
$^3/_8$ inch long, ribbed, and lopsided. From its summit
emerge the petals—2 large ones and 4 smaller ones.
Fields and dry, disturbed sites. Summer–fall.

Melastomataceae, Melastoma Family

Rhexia virginica, Meadow Beauty, Deergrass (plate 319). Our species of Meadow Beauty are erect herbs about 18 inches high, with opposite leaves and attractive 4-petaled flowers 1 inch wide in terminal cymes. The 8 stamens, which have long, curved, bright yellow anthers, are an unusual feature. *R. virginica* has magenta to deep rose flowers and leaves that are at least one-third as wide as long. The stem is square, with equal faces, and has conspicuous flanges or "wings."

Pale Meadow Beauty, **R. mariana,** has pale pink or lavender to nearly white petals and leaves that are no more than one-fourth as wide as long. The stem is winged; its faces are unequal, one pair being broader and convex and the other narrower and concave. Both species are found in damp, usually sandy habitats. Spring–summer.

Onagraceae, Evening Primrose Family

Oenothera spp., Evening Primrose. In this genus the hypanthium is prolonged into a tube beyond the summit of the inferior ovary. The floral parts are in fours, except for the stamens of which there are eight. The stigmas in our species are 4-branched, forming a cross.

Oenothera speciosa, Showy Evening Primrose (plate 320), is the only one occurring in our region with white to pink flowers (the others are yellow-flowered); they are 2 inches or more across. The leaves are elliptic and irregularly blunt-toothed or lobed. The 1- to 2-foot stems sometimes spread widely, forming masses of blooms along roadsides or in fields. Spring–summer.

O. biennis, Common Evening Primrose (plate 321). This species and the next two are among the true "evening primroses," opening late in the afternoon and not wilting until the next day. *O. biennis* is probably the best known of these. Its erect stem may be as tall as 6 feet and is seldom branched except near the top; it has many simple leaves and is often suffused with red. The flowers are fragrant, in a terminal raceme, with petals $^5/_8$ to 1 inch long. In this and the next two species, the capsules are nearly cylindrical in cross-section, not angled. Open areas. Summer–fall.

Evening Primroses are visited during the night by rosy maple moths (Dryocampa rubicunda), *and these pretty little pink-and-yellow creatures can often be seen the next morning, asleep inside a still-open flower.*

O. laciniata, Cut-leaved Evening Primrose. A much shorter plant, frequently branched from the base. Its leaves are pinnately but irregularly lobed. The flowers are solitary from the axils, yellow but soon turning pinkish, with petals less than 1 inch long. Dry soil. Spring–summer.

O. argillicola, Shale Evening Primrose, also is multiple-stemmed. Its cauline leaves are lanceolate with a few obscure teeth. The flowers are showy, with petals from 1 to 1 $^1/_2$ inches long. Mostly on shale barrens. Summer–fall.

O. fruticosa, Sundrops (plate 322). The remaining species have flowers that open in the morning. O. *fruticosa* has handsome, bright yellow flowers with petals up to 1 inch long, and this has led to its frequent cultivation. The capsules are obovoid and strongly 4-winged. This is a highly variable species; plants having oblong capsules and exhibiting minor differences in hairiness have been segregated as **ssp. glauca** (or **O. tetragona**). Woods, meadows, and roadsides. Spring–summer.

O. perennis, Small Sundrops. This is nearly always unbranched. The inflorescence is lax, the buds nodding until each opens in turn. The petals are under $^3/_8$ inch in length, the leaves entire and less than 2 inches long, the principal stem leaves $^1/_8$ to $^3/_8$ inches wide, and the capsule club-shaped. Open areas. Late spring–summer.

O. linifolia is rare in our area, and is unusual for its numerous filiform leaves, which are less than $^1/_{16}$ inch wide. Its flowers are about $^1/_4$ inch across. Dry, sandy sites. Late spring–early summer.

Ludwigia decurrens *(Jussiaea decurrens)*, Primrose Willow (plate 323). *Ludwigia* differs from *Oenothera* in that the hypanthium does not surpass the ovary. The petals are yellow, and all except the last species have alternate leaves. *L. decurrens* is erect, with sessile, narrow lanceolate leaves with edges that extend down-

ward making the stem 4-winged. The flowers are about
³/₄ inch wide, with 4 petals and 8 stamens. Wet areas,
shallow water. Summer–fall.

L. alternifolia, Seedbox, False Loosestrife (plate 324).
This species is common in a variety of wet places, growing up to 4 feet tall. Its leaves are up to 4 inches long
and have tapering bases; they are not decurrent on the
stem. The flowers are about ¹/₂ inch across, with conspicuous spreading, ovate sepals. This and the next species have only 4 stamens.

Much less frequent in the mountains is **L. hirtella,** in
which the leaves are rounded at the base. The narrow
sepals are shorter than the petals, which are larger than
in *L. alternifolia.*

L. palustris, Marsh Purslane. This species is prostrate
or floating, with reddish stems and opposite, ovate leaves
up to 1 inch on long petioles. The minute flowers are
sessile in the axils and lack petals.

Epilobium angustifolium, Fireweed (plate 325). Our
species of *Epilobium* have pink or purplish flowers with
4 petals and 8 stamens, blooming in summer to early
fall. The receptacles and capsules are long and very
slender. Often abundant in recently burned-over areas, *E. angustifolium* is erect, 6 feet or taller, with alternate sessile, lanceolate leaves and long terminal
racemes. The flowers, which are pendent in bud, are
pink to magenta, 1 inch across, the petals rounded
at the apex. The stigma is 4-lobed.

E. coloratum, Purple-leaved Willow Herb (plate 326).
This plant is only half as high as the preceding species
and is often purple-tinged. The leaves are mostly opposite, narrowly lanceolate and fine-toothed, their bases
decurrent on the stem. Its flowers are less than 1/4 inch
wide, with pink or white notched petals and an entire
stigma, but the capsules may be 2 inches long. The
coma—a terminal tuft of hairs on the seeds—is brown.
E. ciliatum *(E. adenocaulon)* is similar but has a whitish coma. Moist, open areas.

E. leptophyllum, Narrow-leaved Willow Herb. This is
a downy plant with flowers like the preceding but slightly

larger. The leaves, however, are linear (less than $\frac{1}{8}$ inch wide) with revolute margins, and are not decurrent. Wet places.

Circaea lutetiana* ssp. *canadensis (*C. quadrisulcata* var. *canadensis*), Enchanter's Nightshade. Enchanter's Nightshades are unique in having their floral parts (petals, sepals, and stamens) in twos, but this is not likely to be noticed except under a hand lens, for the petals—which are white and deeply notched—are less than $\frac{1}{8}$ inch long. The little flowers are numerous in racemes above opposite, petioled, ovate 2- to 4-inch leaves with rounded bases and small distant teeth on the margins. The overall height of this species usually exceeds 1 foot. Rich woods. Summer.

Smaller Enchanter's Nightshade, **C. *alpina*** (plate 327), is well under 1 foot high, and its leaves are less than 2 inches long but have larger coarse, sharp teeth and often are cordate at the base. The flowers are fewer and even smaller than those of the preceding species. Forests at high elevations. Summer–fall.

***Gaura biennis*,** Biennial Gaura, Morning Honeysuckle (plate 328). This is a tall plant (5 feet or higher) with narrow lanceolate leaves and long, branched spikes bearing many small flowers with a tubular hypanthium. The 4 narrow, slightly asymmetric petals are 1/4 inch long, white but soon turning pink. Somewhat deflexed are 8 long stamens and a style with a cross-shaped stigma. Roadsides, fields, and waste places. Summer–fall.

ARALIACEAE, GINSENG FAMILY

***Panax quinquefolius*,** Ginseng (plate 329), is much sought because of the supposed medicinal properties of its root, especially in the Orient. It grows to a maximum height of about 2 feet and has a whorl of 3 long-stalked leaves. Each of these is palmately divided into 5 (occasionally more) stalked, obovate leaflets, the basal pair much smaller than the others. The flowers are small and greenish, in a rounded umbel. Fruits are flattened red berries. Rich woods. Late spring–early summer.

The supply of Ginseng in the Far East was so depleted by thousands of years of harvesting that China found it necessary to begin importing it from North America in the mid-1700s. Today, digging the roots of "sang" in remote southern mountain coves for the Oriental markets is still a lucrative occupation.

P. trifolius, Dwarf Ginseng (plate 330), is only 8 inches high at most. It also has a whorl of 3 leaves, but each usually has only 3 leaflets, and these are elliptic and sessile. Its flowers are white and, as in *P. quinquefolius*, may be either perfect or unisexual. The berries are yellow. Rich woods. Spring.

Aralia nudicaulis, Wild Sarsaparilla (plate 331). This smooth plant has a solitary, ternate basal leaf with 5 pinnate segments in each division. The naked flower stalk is about 8 inches high—much shorter than the leaves—and bears 3 umbels of small greenish white flowers. Woods. Spring–summer.

A. hispida, Bristly Sarsaparilla, is well named, for its lower stem is beset with many slender bristles; there also are several bipinnate cauline leaves. The inflorescence is a loose, open terminal arrangement of long-stalked umbels of small white flowers. Dry, open places. Summer.

A. racemosa, Spikenard (plate 332), is a large, widely branched plant up to 8 feet tall with several compound leaves that may be as much as 2 feet long; the leaflets are heart-shaped at the base. The inflorescence is a large compound panicle made up of many umbels of small white flowers. (This smooth herbaceous plant should not be confused with *A. spinosa*, known as Hercules' Club or Devil's Walking Stick, which is a woody shrub and is armed with stout thorns.) Rich woods. Summer.

APIACEAE, CARROT FAMILY

The plants in this family include not only distinctively flavored culinary herbs such as dill, anise, and caraway and other edibles like celery and parsnip, but some of the most poisonous plants known. They have their flowers in umbels (which are compound in most but simple in *Hydrocotyle* and *Sanicula*) or rarely in heads (*Eryngium*). Frequently an umbel will be subtended by an involucre of bracts; if compound, the individual components are called umbellets, involucels, and bractlets. The leaves may be extremely variable even on a single plant; the forms indicated below for the various genera are the ones usually assumed by the principal cauline leaves in our species.

> Leaves simple: *Hydrocotyle, Bupleurum, Eryngium*
>
> Leaves once compound: *Sanicula, Heracleum, Cryptotaenia, Pastinaca, Oxypolis*
>
> Leaves more than once compound, finely cut: *Daucus, Conium, Chaerophyllum, Anthriscus, Torilis, Foeniculum*
>
> Leaves more than once compound, not finely cut: *Osmorhiza, Angelica, Ligusticum, Cicuta, Taenidia*
>
> Leaves of more than one type: *Zizia, Thaspium*

Leaves simple

Hydrocotyle americana, Pennywort (plate 333). Of all the members of the Carrot Family that occur in our region, only two genera have nearly circular leaves. In our species of Pennywort, they are between 1 and 2 inches in diameter with crenate margins. *H. americana* is terrestrial, and the stems are erect even though very slender. The tiny greenish white flowers are few, in virtually sessile umbels. Wet places. Summer–fall.

H. ranunculoides is a fleshy semi-aquatic with creeping or floating stems, and the umbels are on long peduncles. Its leaves are lobed as well as crenate. Spring–summer.

Bupleurum rotundifolium, Thoroughwax, Hare's Ear. The upper leaves of this plant are roundly ovate, 1 to 2 inches across, entire and perfoliate. There are several compound umbels of small yellow flowers with conspicuous ovate bractlets about $^3/_8$ inch long. Road banks and waste places. Summer.

Eryngium yuccifolium, Rattlesnake Master, Button Snakeroot (plate 334). Strangely for plants that belong to a family also known as the Umbelliferae, the flowers of the Eryngoes are not in umbels. In fact, *E. yuccifolium* is even more of an anomaly in having leaves that are elongate and parallel-veined, which are characteristics of the monocots. This is an erect plant 3 feet or much taller, with stiff, swordlike, spiny-margined basal leaves that are indeed like some Yuccas. It is branched above, with numerous $^1/_2$- to $^3/_4$-inch globose heads crowded with tiny white or greenish flowers that are mostly concealed by stiff bractlets. Roadsides and open, dry woods. Summer.

E. integrifolium is also erect but smaller and has lanceolate to ovate stalked leaves that are net-veined and have toothed or scalloped margins. The flowers are bluish in $^1/_4$-inch roundish heads. Wet fields and mesic woods. Summer–fall.

E. prostratum is a creeping species with ovate to lanceolate leaves very irregularly toothed or lobed. The flowers are blue, not hidden by the bractlets, and are in $^1/_4$-inch cylindrical heads on slender stalks solitary in the leaf axils. Wet or muddy soil. Summer–fall.

Leaves once compound

Sanicula spp., Sanicle. The Sanicles are unusual for having leaves palmately divided almost to the base. There are 3 or 5 segments, but the lateral ones often are deeply cleft, creating the appearance of more. Each of the small, dense umbels contains 3 perfect flowers along with several staminate flowers. These, as well as the fruits, which are covered with hooked bristles, must be observed closely for clues to the identity of the respective species. Woodlands. Spring–summer.

> Styles longer than bristles; staminate flowers numerous:
> S. *gregaria*, S. *marilandica*
> Styles shorter than bristles; staminate flowers few or absent:
> S. *trifoliata*, S. *canadensis*, S. *smallii*

Sanicula gregaria, Clustered Snakeroot (plate 335), has 5- or sometimes 3-parted leaves. Its yellow petals serve to separate it from all the others.

S. marilandica, Black Snakeroot, has 5-parted leaves but the lowermost pair are often cleft so that they appear 7-parted. Its flowers are greenish white.

S. trifoliata has 3-parted leaves, the lateral segments sometimes incised but not deeply. The flowers are white.

The fruit is sessile, and its bristles are exceeded by both the long-stalked staminate flowers and the persistent beaklike sepals of the pistillate flowers.

In **S. canadensis** and **S. smallii**—both white-flowered species—the leaves are 3- or 5-parted, and neither the staminate flowers nor the sepals appreciably surpass the bristles. The fruits are short-stalked in the first species, sessile in the second.

Heracleum maximum (*H. lanatum*), Cow Parsnip (plate 336). This is a tall, coarse plant sometimes reaching 9 feet in height, hairy below, with ternately or pinnately compound leaves. The sheaths of the petioles are expanded to as much as 2 inches wide. The umbels are up to 8 inches across; the flowers have white, somewhat irregular corollas, the outer ones larger with lobed petals. Roadsides, meadows, and moist places. Summer.

A much smaller, more delicate plant, **Cryptotaenia canadensis,** or Honewort, has minute white flowers in loose, irregular umbels made up of several primary rays of unequal length and few-flowered umbellets. Its leaves are ternately divided into ovate, toothed, and occasionally lobed segments; the sheaths on the stalks are not dilated. Moist woods. Spring–early summer.

Pastinaca sativa, Wild Parsnip (plate 337), is a tall plant, up to 6 feet. Its leaves are long-stalked and pinnately divided, the segments toothed throughout. The flowers are yellow, in umbels up to 6 inches wide. Waste places. Summer.

Oxypolis rigidior, Cowbane (plate 338), is a smooth plant of wet places, blooming in midsummer and fall. Its leaves are once-pinnate; the leaflets may be entire or bear large, sharp teeth, and may vary from $1/4$ to $1 1/2$ inches in width. The white flowers are borne in small, dense, hemispheric umbellets which are well separated from each other by the widely spreading primary rays.

Leaves more than once compound, finely cut

Daucus carota, Wild Carrot, Queen Anne's Lace (plate 339). An ancestor of the familiar cultivated carrot, this is one of the most common weeds of fields and waste areas. It is usually branched, and the leaves are pinnately decompound with all of the segments very narrow. The umbels are convex at first, then flat, the terminal ones 2 ¹/₂ inches or more across, densely white-flowered but with a solitary dark purple flower in the center. The bracts beneath the umbels are up to 1 ¹/₂ inches long and pinnately divided into filiform segments. Summer–fall.

There is a legend behind the naming of Wild Carrot as "Queen Anne's Lace." It says that while she was making a doily she pricked her finger and spilled a drop of royal blood in the center of her handiwork.

D. pusillus is smaller and typically unbranched. Its terminal umbels do not exceed 2 inches in width, and there is no colored central flower. The bracts are ³/₄ inch or less. Spring.

Conium maculatum, is the Poison Hemlock that was used to put Socrates to death. It is a freely branched plant, 6 feet or more high, with a smooth, purple-spotted stem and pinnately decompound leaves with expanded petiolar sheaths. Its white-flowered umbels are about 2 inches wide, and the bracts simple, lanceolate, and less than ¹/₄ inch long. A dangerous weed growing in many kinds of waste places. Spring–summer.

Chaerophyllum tainturieri, Wild Chervil (plate 340), is a hairy, erect plant, branched above. It has leaves that are ternately decompound, the divisions pinnately divided and redivided. There are only 1 to 3 rays; each umbellet contains a few tiny white flowers and is subtended by leaflike bractlets. Waste places. Spring. **C. procumbens,** Spreading Chervil, is smooth and decumbent, branched from the base.

Anthriscus sylvestris, Cow Chervil, is a large, freely branched plant with twice or thrice pinnately compound leaves, the segments deeply cut. The umbels have 6 to 10 primary rays up to 1 ¹/₂ inches long. The white flowers have 5 obovate petals. Disturbed ground. Summer. **A. scandicina** is a rare escape. It has 5 or fewer principal rays.

Torilis arvensis (*T. japonica*), Hedge Parsley, has leaves that are pinnately (or the upper ones ternately) decompound, the segments narrow and sharply toothed. The umbels bear a few small white flowers and are on long stalks. Waste areas. Summer.

Foeniculum vulgare, Fennel. This well-known culinary herb has become naturalized at a number of places in the South. Its leaves are repeatedly and pinnately dissected into filiform segments. The umbels are up to 4 inches wide, yellow-flowered, without involucres. Summer.

Leaves more than once compound; not finely cut

The leaves of **Osmorhiza claytonii,** Sweet Cicely, are mostly twice-ternately compound. There are only 3 to 6 primary rays, and a few small white flowers in each umbellet. The style, including its swollen base (the stylopodium) is shorter than the petals and less than $^1/_{16}$ inch in fruit. In contrast to this hairy species, **O. longistylis,** Aniseroot (plate 341), is essentially smooth. Its style and stylopodium are longer than the petals and about $^1/_8$ inch in fruit. Woods. Spring.

Angelica triquinata, Filmy Angelica (plate 342). These are large plants with pinnately decompound leaves, their petioles greatly expanded at the base; the leaves are reduced in size above and the upper ones sometimes consist only of a sheath. A. *triquinata* has a stout, smooth, sometimes purplish stem, and umbels 3 to 6 inches wide. The flowers are white, often tinged with green or pink. The leaflets are coarsely toothed with minutely ciliate margins and are sharp-pointed at the apex. High elevation woods and balds. Summer–fall.

A. atropurpurea, Purple Angelica, is much less common. It has a stout, smooth, dark purple stem and larger umbels with as many as 45 rays. The sharp-pointed leaflets are eciliate but have narrow, pale margins. Wet places. Spring–early summer.

A. venenosa, Hairy Angelica, differs from the other two in several respects. It has a slender stem, densely fine-hairy above and in the inflorescence. The leaves are finely toothed and blunt at the apex. The flowers are white. Dry, open areas. Summer.

Ligusticum canadense, American Lovage (plate 343), is a tall plant (up to 5 feet) with leaves ternately or pinnately decompound, the segments ovate and sharp-toothed; the petioles have narrow basal sheaths. The main stem sometimes ends in a whorl of branches, making the terminal umbel appear doubly compound. The flowers are white. Rich woods. Spring–summer.

Cicuta maculata, Water Hemlock, Spotted Cowbane (plate 344). This is a tall, branching plant with a very poisonous root, common in wet places and flowering from summer to fall. The stem is smooth, sometimes streaked with purple, and the leaves are pinnately decompound with narrowly lanceolate, coarsely toothed leaflets tapering to a point. It bears large, many-rayed umbels of white flowers, with very few or no involucral bracts.

Taenidia integerrima, Yellow Pimpernel (plate 345). A tall, glaucous-stemmed yellow-flowered species, this has decompound foliage with ovate to elliptic leaflets that have consistently entire margins—an uncommon feature in the Apiaceae. The umbel is open, its slender primary rays up to 3 inches long, but unequal. Dry slopes and open woods. Spring–early summer.

Leaves of more than one type

Zizia spp., Golden Alexanders. All species of *Zizia* have bright yellow flowers and bloom in woods in spring. Some bear a closer resemblance to certain plants in *Thaspium* than to some others in their own genus, and vice versa. The confusion between genera can be avoided, however, by remembering that in *Zizia* the central flower of each umbellet is sessile, while in *Thaspium* all are stalked.

In **Zizia aptera,** Heart-leaf Alexanders (plate 346), the basal leaves are simple, have a cordate base, and are long-petioled; the cauline leaves are once- or twice-ternate.

Z. aurea and **Z. trifoliata,** both known as Golden Alexanders, resemble each other but differ from the preceding by having their lower leaves twice-ternately compound and the upper ones once-ternate. In Z. *aurea*, however, the leaflets are finely toothed (10 to 25 teeth per inch of margin), while Z. *trifoliata* is coarsely toothed (5 to 10 per inch). Also, the first species has 10 or more primary rays, the second usually fewer.

Thaspium trifoliatum, Meadow Parsnip (plate 347). With its dark purple flowers, this species of *Thaspium* would seem safe from becoming confused with any of the *Zizias*—until it is realized that there is an even more widespread form, **T. trifoliatum var. flavum,** which has yellow flowers. In either case, the basal leaves may be either simple and ovate with a cordate base or once-ternate; the cauline leaves are pinnate with 3 or 5 leaflets.

T. barbinode, Hairy-jointed Meadow Parsnip (plate 348). Small hairs at the upper nodes are a field mark of this yellow-flowered species. The basal and principal cauline leaves are twice pinnately compound or ternately compound, the ultimate segments ovate to lanceolate.

T. pinnatifidum has decompound leaves with the ultimate divisions linear or nearly so, none more than $1/8$ inch wide. Its flowers are creamy white.

CORNACEAE, DOGWOOD FAMILY

Cornus canadensis, Bunchberry (plate 349). Like those of the related Flowering Dogwood tree, Bunchberry blossoms are made conspicuous by four large white bracts surrounding a central cluster of 4-petaled yellowish flowers. Fruits are bright red drupes. Damp woods at high elevations. Early summer.

ERICACEAE, HEATH FAMILY

Chimaphila maculata, Spotted Wintergreen (plate 350). This species and the next have cauline leaves and clusters of white or pink $3/4$-inch flowers with 5 spreading petals and 10 stamens each with a pair of conspicuous tubular anthers. C. *maculata* is instantly recognizable by its evergreen leaves, which are dark green with broad, whitish stripes along the veins, are widest near the base, and have large, distant teeth along the margins.

C. umbellata, Pipsissewa, Prince's Pine. The leaves of *C. umbellata* are uniformly green and shiny, without lighter markings, are widest near the apex, and have small teeth. It is more northern in distribution than *C. maculata.* Both are woodland species flowering in late spring and early summer.

Pyrola rotundifolia var. americana (*P. americana*), Round-leaved Pyrola (plate 351). There are many differences between Round-leaved Pyrola and *Chimaphila*: Its leaves are all basal and long-stalked, the flowers are borne in a raceme instead of terminally, and the petals do not spread as widely. A conspicuous feature is the style, which is curved and extends well beyond the corolla. Woodlands. Summer.

Epigaea repens, Trailing Arbutus (plate 352). Smaller than most of our other plants in the Heath Family, this is one of the most admired and sought after. Its leathery oblong leaves are on creeping stems and often conceal the little spikes of fragrant $^1/_2$-inch white or pink salverform flowers. Dry woods and road banks. Early spring.

Gaultheria procumbens, Wintergreen, Teaberry, Checkerberry (plate 353). Unlike Trailing Arbutus, Wintergreen is an erect plant with a few smooth leaves crowded near the top. Several white, barrel-shaped blossoms less than $^3/_8$ inch long are borne on short pedicels. The red berries often persist throughout the winter. Woods, flowering in summer.

These little plants contain small amounts of oil of wintergreen (methyl salicylate)—but enough for them to have served as the primary source until it was discovered that it could be distilled more economically from birch bark, and still later that it could be synthesized.

Monotropa uniflora, Indian Pipe, Ghost Flower (plate 354). The two species of *Monotropa* are common woodland saprophytic plants with stems about 6 to 8 inches high, scalelike leaves, and flowers with 4 or 5 separate petals, all parts colored alike. Indian Pipe is waxy and translucent, white or pale pink, with a solitary nodding flower $^3/_4$ inch long. Summer and fall.

A Cherokee legend tells how the chiefs of several quarreling tribes met in council, smoking the pipe for seven days and nights, but failed in the end to settle their dispute. This displeased the Great Spirit, who had decreed that the pipe was to be smoked only after peace had been achieved, so to make an example of them he transformed the old men with their bowed heads into the grayish flowers we call Indian Pipes.

M. hypopithys, Pinesap (plate 355), is similar but bears several smaller ($^1/_2$-inch) flowers in a curved raceme. Its color may range from yellow or tawny to pink or reddish. Summer and fall.

Monotropsis odorata, Sweet Pinesap, is another saprophyte but much smaller—not more than 4 inches high and with corollas only $^1/_4$ inch long. The flowers are campanulate with united petals, vary in color from pinkish to purplish brown, and are fragrant. Rare, but also easily overlooked, it usually is found under pines. Most plants flower in early spring, some in fall.

DIAPENSIACEAE, DIAPENSIA FAMILY

Galax aphylla (G. *rotundifolia,* G. *urceolata),* Galax, Beetleweed (plate 356). Galax has a cluster of long-stalked basal evergreen leaves, nearly circular with palmate veins, sharp-toothed, glossy, and often aging to wine-red or bronze. Surpassing these to a height of 1 to 2 feet is a dense, narrow, spikelike raceme of flowers with white 5-lobed corollas less than $^1/_4$ inch long. Rich woods and forest margins. Late spring–early summer.

Gathering the attractive and durable leaves of Galax for sale, to be used in floral arrangements and especially as Christmas decorations, was for years a source of income for many mountain families.

Shortia galacifolia, Shortia, Oconee Bells (plate 357). Sharing the Diapensia Family but quite different is Shortia, a southern endemic now rare in the wild. Its leaves are widely elliptic with pinnate veins, toothed, and shiny. The flowers are campanulate, $^3/_4$ to 1 inch long, with 5 white petals blunt-toothed at the apex, and are solitary on scapes up to 6 inches long. On wooded stream banks. Early spring.

PRIMULACEAE, PRIMROSE FAMILY

Lysimachia quadrifolia, Whorled Loosestrife (plate 358). Our species of *Lysimachia* have simple leaves, paired (but whorled in a few instances), and bloom in summer. Their flowers have 5 yellow petals joined at the base. *L. quadrifolia* has lanceolate leaves in regularly spaced whorls of 4. Its star-shaped flowers are about $^1/_2$ inch across, marked with red in the center, and are borne on long, slender stalks, one in the axil of each of the median leaves. Open areas.

L. nummularia, Moneywort (plate 359). This is the only creeping species in our region. Its $^1/_2$- to 1-inch leaves are nearly circular and short-stalked. The 1 inch wide flowers are solitary in the axils. Lawns and moist ground.

L. terrestris, Swamp Candles (plate 360). Only two of our species have terminal inflorescences in which the bracts subtending the pedicels are much smaller than the stem leaves. *L. terrestris*, which may grow to 3 feet high, has flowers in an erect, slender raceme up to 1 foot long, with many $^1/_2$-inch flowers, marked with dark lines. Wet areas. The other, **L. fraseri,** is much larger and has whorled leaves, but its flowers are about the same size. They are borne in a panicle. Meadows and roadsides.

L. ciliata (*Steironema ciliatum*), Fringed Loosestrife. Our remaining species (which formerly were placed in the genus *Steironema*) have stalked flowers arising singly from the axils of the leaves, which are only gradually reduced in size toward the summit. The petals are unmarked and usually have a single tooth or are fringed at the apex. *L. ciliata* has ovate-lanceolate leaves on petioles that are ciliate over their entire length. The 1-inch nodding flowers have ovate petals. Wet places. Summer–fall.

L. tonsa (*S. intermedium*) is similar but its petioles are not fringed and it grows in dry woods.

True to its name, **L. lanceolata** (*S. lanceolatum*), Lance-leaf Loosestrife (plate 361), has leaves that are tapered at both ends, without a differentiated petiole. It is a stoloniferous plant, and has $^3/_4$-inch flowers. The similar **L. hybrida**—which is not a hybrid—is not stoloniferous, and its leaves have short but distinct stalks. Wet places.

L. quadriflora (*S. quadriflorum*), Prairie Loosestrife. This can readily be distinguished from the others by its stiff, ascending linear leaves, which are only about $^1/_8$ inch wide. The flowers are $^3/_4$ inch across and are mostly concentrated in the upper part of the plant. Wet soil.

Anagallis arvensis, Scarlet Pimpernel, Poor Man's Weatherglass (plate 362), is a low, weakly sprawling weed with pairs of 1-inch sessile, ovate leaves. Its flowers are solitary on slender stalks in the axils; the 5-lobed corolla is $^1/_4$ inch wide and scarlet (or sometimes blue). They open only in sunny weather, closing with the approach of clouds. Waste areas and lawns. Summer.

Trientalis borealis, Star Flower (plate 363), has a single terminal whorl of 5 to 9 lanceolate leaves, surmounted by one to a few white flowers on slender but erect pedicels. The rotate corollas are $^1/_2$ inch wide, each with 7 (more or less) pointed lobes which give it a starlike appearance. Moist woodlands. Late spring–early summer.

Dodecatheon meadia, Shooting Star (plate 364). The flower structure of *Dodecatheon* is unusual, and this is our only species. The leaves are oblanceolate and up to 10 inches long, all basal. The 18-inch scape is topped by an umbel of several nodding flowers with their 5 corolla lobes swept back to expose the stamens, which are joined into a conical beak. The petals may be white or pink. Rich woods, often calcareous. Spring.

The flower color in D. meadia is variable, but white corollas seem to be more prevalent in the southern mountains than to the north. At one time white-flowered plants were segregated as a separate species, D. hugeri, but that distinction is no longer recognized.

Samolus parviflorus (*S. floribundus*), Water Pimpernel, Brookweed. This is a plant of wet habitats, usually less than 1 foot high, with untoothed spatulate leaves about 2 inches long, alternate but with a basal cluster. Its flowers are white, deeply 5-lobed, only $^1/_8$ inch across, and are borne on slender stalks in a loose terminal raceme. Late spring–fall.

LOGANIACEAE, LOGANIA FAMILY

Spigelia marilandica, Indian Pink, Pinkroot (plate 365). This striking native wildflower is borne on a plant up to 2 feet tall with opposite, sessile leaves averaging 3 inches in length. The flowers are tubular, about 1 $^1/_2$ inches long, crimson on the outside and yellow within. The latter color is revealed as the 5 lobes of each corolla spread to create a yellow star at the summit. The flowers are closely arranged in a one-sided cyme. Rich woods. Late spring–early summer.

A more modest member of the family is **Polypremum procumbens,** a low, well-branched, wiry plant with opposite linear leaves up to $^3/_4$ inch long and 4-lobed white flowers a mere $^1/_8$ inch across. Roadsides and waste places. Summer–fall.

GENTIANACEAE, GENTIAN FAMILY

Gentianopsis crinita (*Gentiana crinita*), Fringed Gentian (plate 366). Our Gentians are late summer to fall blooming plants, usually preferring moist habitats, with blue or violet flowers (although greenish or even white ones can be expected) and opposite entire leaves. For the most part, they have 5-lobed corollas and calyxes, but this genus provides a notable exception. Not only are its 1 $^1/_2$- to 2-inch flowers 4-merous, but the corolla is deeply fringed—a feature that has helped to make it one of our most admired wildflowers. Moist woods and banks. Fall.

Gentianella quinquefolia (*Gentiana quinquefolia*), Stiff Gentian, Agueweed (plate 367). Stiff Gentian is instantly recognizable by its very numerous flowers in dense terminal and axillary cymes. They are tubular, $^3/_4$ inch long, and vary greatly in color from deep blue or purple to lilac or white. The plant is well-branched and has a sharply angled stem. Open areas.

Gentiana clausa, Closed or Bottle Gentian (plate 368). Our species of *Gentiana* have flowers 1 to 1 $^1/_2$ inch long. In many cases these can be identified only by giving careful attention to small details—such as the pleats that alternate with and are connected to the 5 corolla lobes. In our area, G. *clausa* is the most common of the "closed" Gentians, in which the deep blue or purple flower opens only when forced by a pollinating insect, then immediately closes after it. The pleats have 2 or 3 teeth at their summits and are about as long as, but are hidden by, the lobes of the closed corolla.

Much less frequent in the South is **G. *andrewsii*,** in which the pleats end in fringes which protrude beyond the shorter corolla, forming a nearly truncate summit.

G. *saponaria*, Soapwort Gentian. In this species the flowers are often paler than the last two but with dark lines within, and may be closed or slightly open. The corolla lobes are not much longer than the pleats, which are incised. The calyx lobes are narrow and $^3/_8$ inch long.

Unlike our other species, which are smooth, Striped Gentian, **G. *decora*** (plate 369), is a minutely pubescent plant. Its flowers are open, and the corolla lobes conspicuously exceed the broad, erose pleats. They are whitish or blue with darker blue or violet stripes and are in a dense terminal cluster. The calyx lobes are narrow and less than $^1/_4$ inch long.

Notwithstanding its specific epithet, **G. *villosa*,** also called Sampson's Snakeroot, is smooth. The partially open corolla is greenish or yellowish white with purple stripes within, and the calyx lobes are linear and unequal but averaging $^3/_4$ inch long.

G. *linearis*, Narrow-leaved Gentian, has distinctive linear-lanceolate leaves less than $^3/_8$ inch wide. It has only a few flowers in a terminal cluster and the upper axils.

***Obolaria virginica*,** Pennywort (plate 370). This is a fleshy plant only about 4 inches high with opposite leaves, which are scalelike below and obovate and purple-tinged above. The flowers are terminal (solitary) and axillary (clusters of 1 to 3), funnel-shaped with 4 pointed lobes, $^1/_2$ inch long, white or bluish. The calyx has 2 foliaceous lobes. The small size of this plant and its dull coloring make it easy to overlook. Woodlands. Spring.

Frasera carolinensis (*Swertia caroliniensis*), American Columbo (plate 371). Monument Plant is another and especially fitting name for this stately plant, which may exceed a height of 6 feet and terminates in a pyramidal panicle of flowers one-third as tall. The leaves are lanceolate and mostly in whorls of 4, up to 15 inches long at the base and progressively reduced upward. The 4-merous rotate flowers are pale greenish white or yellow dotted with brown-purple; each petal is ⅝ inch long and bears near the middle a prominent green gland fringed with long hairs. Meadows and open woods, often in calcareous soil. Late spring–early summer.

Sabatia angularis, Rose Pink, Bitterbloom (plate 372). Our species of *Sabatia* are up to 3 feet tall and have beautiful pink flowers with a rotate corolla about 1 inch wide. Inside at the base there is a star-shaped "eye," usually yellow or greenish with a red border. *S. angularis* is so-named because its stem is 4-angled and sharply winged. The upper branches are opposite, as are its sessile leaves, which are broad and rounded at the base. Moist fields and roadsides. Summer.

S. brachiata also has opposite branches but the stem is not winged. Its leaves are narrow, even linear, and tapered at the base. Dry areas.

Slender Marsh Pink, ***S. campanulata,*** differs from these by having its upper branches diverge alternately. Damp fields and woodlands.

Menyanthes trifoliata, Buckbean (plate 373). Circumpolar in distribution, Buckbean barely enters our region. It is a smooth rhizomatous plant with long-petioled trifoliolate leaves arising from the base, the leaflets oblong to ovate, entire, up to 3 inches long. The flowers are ½ inch across with 5 spreading lobes, white with copious hairs within, in a raceme. In shallow water, bogs, and marshes. Spring–summer.

Apocynaceae, Dogbane Family

Apocynum androsaemifolium, Spreading Dogbane (plate 374). The Dogbanes are well-branched plants with milky latex, opposite leaves, and clusters of small flowers terminating the main stem and branches. They usually are found in dry soils, often in fields or disturbed sites, blooming in summer. Spreading Dogbane has nodding, bell-shaped flowers with 5 recurved lobes, $^1/_4$ to $^3/_8$ inch long, pink with deep rose striping within.

A. cannabinum, known as Indian Hemp, is more erect and has clusters of smaller, erect, cylindrical, greenish white flowers without stripes. The terminal inflorescence is often surpassed by lower lateral branches. The leaves are rounded at the base, have distinct petioles, and are spreading or ascending.

Plants are often seen which possess some of the characters ascribed to each of the preceding species but do not completely conform to either. These have been aptly called Intermediate Dogbane and variously construed as hybrids or as belonging to a separate species, **A. medium.**

A. sibiricum, Clasping-leaved Dogbane, has sessile, often cordate-based and auriculate leaves. Its flowers are white, erect, and not much more than $^1/_8$ inch long.

Plants can often be recognized as Apocynums *by the presence of dogbane beetles* (Chrysochus auratus)—*beautifully iridescent green and bronze insects that apparently feed on nothing but Dogbanes and Milkweeds. Also distinctive are the pairs of long, slender seed pods, which may be up to 15 times as long as an individual flower.*

Amsonia tabernaemontana, Blue Star, Blue Dogbane (plate 375). This tall, smooth plant contains milky sap (which is characteristic of the family) but has alternate leaves (which are not). The flowers are salverform, $^1/_2$ inch across, with 5 narrow, spreading blue petals, in a crowded terminal cluster; the corolla tube is downy on the outside. Moist woods and stream banks. Spring.

Vinca minor, Periwinkle, Myrtle (plate 376), is an evergreen creeping plant of European origin widely planted as ground cover and often persisting or escaping to the woods. It has opposite shiny leaves and blue-violet flowers 1 inch wide. Open woods, fields, and roadsides. Spring.

V. major is a much less frequent escape. Its leaves differ in having ciliate margins, and the flowers are larger, up to 2 inches.

ASCLEPIADACEAE, MILKWEED FAMILY

Asclepias spp., Milkweed. The anatomy of a Milkweed flower is quite complex. In order to identify individual species, however, it is necessary only to recognize the most conspicuous external parts. They are: 1) the corolla, which is deeply divided into 5 petals, usually reflexed and concealing the calyx; 2) the corona, which consists of 5 curved hoods alternating with the petals and standing more or less erect; and 3) the 5 slender horns partially or entirely enclosed by the hoods (see plate 382 of individual A. *exaltata* floret).

Most of our Milkweeds are erect, unbranched plants with milky sap, simple entire leaves in pairs or whorls, and small flowers in terminal or axillary umbels. They are presented here in the following order according to the predominant color of their flowers:

> Flowers orange: A. *tuberosa*
> Flowers pink, rose, or purple: A. *quadrifolia*, A. *incarnata*,
>   A. *purpurascens*, A. *syriaca*, A. *amplexicaulis*
> Flowers white: A. *variegata*, A. *exaltata*, A. *verticillata*
> Flowers green: A. *viridiflora*

Asclepias tuberosa, Butterfly Weed, Chigger Weed, Pleurisy Root (plate 377). This is unique among our species on several counts: the color of its flowers, which are bright orange (but often tend toward yellow or scarlet); its alternate leaves, which are abundant; and its sap, which is not milky. Its hairy stems often are branched above and bear numerous profusely flowered umbels. Dry fields and roadsides. Late spring–summer.

A. quadrifolia, Four-leaved Milkweed (plate 378). Unlike the others in this color group, A. *quadrifolia* is a slender, delicate plant with thin ovate-lanceolate leaves, opposite except for the median ones, which are larger and appear whorled. Its umbels are loosely flowered. The flowers are pink or nearly white, the horns short, flattened, and strongly curved. Spring–summer.

A. *incarnata,* Swamp Milkweed, has narrow (less than 1 1/4 inch wide) lanceolate leaves and numerous umbels under 2 inches across. Its flowers are deep pink to rose-purple and small, the petals 1/4 inch long, and the hoods half as high with much longer horns. Most of our plants are the densely pubescent and seldom branched **ssp. *pulchra*** (plate 379).

A. *purpurascens,* Purple Milkweed. The leaves of Purple Milkweed are ovate to elliptic and wider (to 3 inches), and the umbels broader and fewer (sometimes only one) than those of A. *incarnata.* The petals are deep rose and the hoods paler and 1/4 inch high, with horns that are much shorter and curved. Dry, open sites. Summer.

A. *syriaca,* Common Milkweed. In Common Milkweed the leaves are thick and widely elliptic, with distinct petioles. Its umbels may be as much as 4 inches across, and are many-flowered, rounded, and often nodding. The flowers are fragrant and vary in color but generally are a dull rose tinged with green. When in fruit, this species can be distinguished from the others by its warty pods. Fields and waste areas. Late spring–summer.

A. *amplexicaulis,* Blunt-leaved Milkweed. The broad, wavy-margined leaves with their clasping bases serve to identify this species. It is a smooth, unbranched plant bearing a solitary terminal umbel of rose-purple flowers sometimes suffused with green. The horns are slender and much exceed the hoods. Dry, open areas. Spring–summer.

A. *variegata,* White Milkweed (plate 380). The flowers of A. *variegata* are bright white except for a purple center and are crowded into one or a few rounded, compact clusters. The petals are flaring, the hoods saclike, and the horns very short. Spring–summer.

A. *exaltata,* Poke Milkweed (plate 381 and plate 382). This tall plant has several loose, few-flowered, drooping umbels. The petals are pale greenish white, the hoods white and toothed at the summit, and the horns conspicuously exserted and incurved. Woodland margins. Late spring–summer.

A. verticillata, Whorled Milkweed (plate 383). This is a slender, erect plant clothed from top to bottom in whorls of 3 to 6 linear leaves only $^1/_{16}$ inch wide. Several umbels from the upper nodes bear small white or greenish white flowers. Open areas. Summer.

A. viridiflora *(Acerates viridiflora)*, Green Milkweed. The flowers of this species are unique not only because of their green color but also the absence of horns from their coronas. The umbels are hemispheric and densely flowered, lateral from the upper nodes, either sessile or peduncled. Dry fields and waste areas. Summer.

Milkweed flowers possess an interesting device for promoting cross-pollination. Their pollen is contained in sacs (pollinia) which are connected in pairs much like tiny saddlebags and located behind narrow slits in the corona. Frequently an insect will catch its foot in one of these slits, and when it pulls loose it unwittingly carries the pollinia away with it.

Matelea carolinensis, Climbing Milkweed (plate 384). Our species of *Matelea* (formerly placed in the genus *Gonolobus* by some authors) are twining vines with heart-shaped leaves and stalked axillary clusters of flowers in summer. In M. *carolinensis* there are fewer than 10 flowers in each inflorescence. The corolla is brownish purple, about $^3/_4$ inch across, with widely spreading elliptic lobes that are glandular-hairy on their outer surface. Woods and moist thickets.

M. obliqua has more flowers in each cluster. As in the preceding species, the corolla lobes have glandular hairs on the outside, but they are linear, $^1/_2$-inch long, and ascending. Calcareous soils.

M. gonocarpa, Angle Pod. In this and the next species, the flower color may vary from greenish to purplish, usually darker at the center. M. *gonocarpa* has narrowly lanceolate corolla lobes $^1/_2$ inch long and smooth on both surfaces. In the similar **M. suberosa** they are only $^3/_8$ inch long but relatively wider, and are finely pubescent on the upper surface. Alluvial woods.

Cynanchum laeve (*Ampelamus albidum*), Blue Vine. This twining vine has long-stalked triangular leaves with a wide, rounded basal sinus. There are stalked clusters of very small white flowers in the leaf axils. The petals are erect, spreading only slightly, about ¼ inch long. Thickets and fields. Summer.

CONVOLVULACEAE, MORNING GLORY FAMILY

Calystegia sepium (*Convolvulus sepium*), Hedge Bindweed (plate 385). With their funnelform flowers the Bindweeds resemble Morning Glories (*Ipomoea*), but have a pair of bracts below the calyx; in the genus *Calystegia* these are large and conceal the sepals. *Calystegia sepium* is a smooth twining or trailing vine and bears white or pink flowers, 1 ½ to 3 inches long, solitary in the axils of its leaves, which are halberd- or arrowhead-shaped with pointed or squared lobes. Thickets, fields, and disturbed areas. Spring–summer.

C. sericata (*Convolvulus sericatus*) can be distinguished by the soft, felty pubescence on the trailing stem and leaves. The few flowers are solitary in the axils, white, about 2 inches long. Open slopes. Summer.

C. spithamaea (*Convolvulus spithamaeus*), Low Bindweed, has an erect stem, only 1 or 2 feet long. Its leaves are oblong, and the flowers, which are borne only in the lower axils, are white and about 2 inches long. Dry rocky or sandy soil. Spring–summer.

Convolvulus arvensis, Field Bindweed (plate 386). The calyxes of Field Bindweed are accompanied by a pair of bracts but—unlike *Calystegia*—they are narrow and well below the flower so they do not conceal the sepals. The leaves are broadly sagittate, 1 to 2 inches long, and the white or pink flowers are only ¾ inch long. Fields and disturbed areas. Spring to fall.

Ipomoea purpurea, Morning Glory (plate 387). The vines in the genus *Ipomoea* can be differentiated from those in *Calystegia* and *Convolvulus* by the absence of bracts beneath the calyx. *I. purpurea* is the common Morning Glory, imported from the American tropics, widely cultivated, and now thoroughly naturalized. Its large (2- to 3-inch-long) funnelform flowers may be purple, blue, pink, white, or variegated. The leaves are broadly heart-shaped. Summer–fall.

I. pandurata, Wild Potato Vine (plate 388), is also called Man-of-the-Earth because of its enormous tuberlike root. Its leaves are heart-shaped with a tapering apex, and the 2- to 3-inch-long flowers are white with a crimson throat. Fencerows and roadsides. Spring–summer.

In Ivy-leaved Morning Glory, *I. hederacea,* the leaves are circular in general outline but deeply 3-lobed and heart-shaped at the base. Its flowers are usually blue or purple, 1 ¹/₂ to 2 inches long. Fields and waste places. Summer to fall.

I. lacunosa, Small White Morning Glory (plate 389), has white flowers less than 1 inch long. It blooms in summer and fall, until frost.

I. coccinea (*Quamoclit coccinea*) (plate 390) is called Red Morning Glory for good reason. Its flowers are bright scarlet and salverform, the narrow tube about 1 inch long and the spreading limb up to ³/₄ inch across. The leaves are ovate with a cordate base and are pointed at the apex. Fields and waste places. Summer–fall.

CUSCUTACEAE, DODDER FAMILY

Cuscuta spp., Dodder, Love Vine (plate 391). These are twining vines that parasitize a variety of host plants by means of small suckers (haustoria); leaves are scale-like or absent. There are several species in our region, but differentiating them requires minute examination of flower parts and the assistance of a more elaborate key than can be provided here. At the generic level, our plants can be recognized by the yellow or orange stems with clusters of small (less than ¹/₄-inch) white or creamy bell-shaped flowers. Summer and fall.

The air of mystery surrounding many parasitic plants has made them good raw material for legends and superstitions. This may help to account for the great number of curative properties attributed to Dodder by Native Americans, who prescribed it for a host of diseases and injuries and, perhaps strangest of all, as a contraceptive.

POLEMONIACEAE, PHLOX FAMILY

Phlox spp., Phlox. Many of the wildflowers in this genus will be recognized by their similarity to garden varieties. They have opposite, entire leaves and corollas in various shades of pink or purple (as well as white), consisting of a narrow tube that terminates in 5 flaring lobes. In the following treatment our plants have been arbitrarily divided between 1) a half dozen discrete species that come into flower during March or April, and 2) a rather confusing group that usually do not bloom until May or June and may continue until frost.

> Early-flowering species: *P. subulata*, *P. nivalis*, *P. stolonifera*,
> *P. amoena*, *P. pilosa*, *P. divaricata*
> Late-flowering species: *P. paniculata*, *P. amplifolia*, *P. maculata*,
> *P. ovata*, *P. buckleyi*, *P. carolina*, *P. glaberrima*

Phlox subulata, Moss Phlox, Moss Pink (plate 392), is a native species, more common in the Northeast; it has been extensively cultivated and often escapes to the wild from wall and rock garden plantings. It is prostrate, mat-forming, and semi-evergreen, with numerous awl-shaped leaves about 1 inch long. The many flowers are $1/2$ to $3/4$ inch wide, pink to rose-purple or white, with a dark eye; the lobes are notched at the summit. The stamens are partly exserted from the corolla tube. Dry, rocky places.

P. nivalis, Trailing Phlox, is a similar but more southern species, though infrequent in the mountains. Its leaves are less than $1/2$ inch long, but the flowers may approach 1 inch in width; the stamens do not protrude. Sandy soil.

P. stolonifera, Creeping Phlox (plate 393). The trailing stems of *P. stolonifera* root at the nodes, sending up leafy sterile shoots as well as stalks bearing few-flowered cymes of rose-purple flowers up to 2 inches across with entire lobes, the stamens slightly protruding. The lower leaves are numerous, short-petioled, and spatulate. Moist woods.

P. amoena, Hairy Phlox. The cymes of *P. amoena* are compact, its flowers on very short pedicels. It is decumbent at the base, with erect sterile as well as fertile shoots.

The corollas are up to ³/₄ inch across and vary as to color, the lobes obovate and entire, and the stamens included. Its leaves are elliptic or linear, less than ¹/₄ inch wide and under 2 inches long, rather blunt at the apex. Dry woods and road banks.

P. pilosa, Downy Phlox. Primarily a lowland species, this differs from *P. amoena* in being taller (to 2 feet or more), usually without sterile shoots, and in having a loosely branched inflorescence. Its leaves are narrowly lanceolate, up to 4 inches long, and long-tapered to a sharp point. Open areas.

Wild Blue Phlox, **P. divaricata** (plate 394), has light blue-violet flowers—a feature that distinguishes it from our other species, which are more pinkish or purplish than blue. The corolla sometimes exceeds 1 inch in width; its lobes are wedge-shaped and typically are notched at the apex. There are only a few pairs of leaves on each flowering stem. Woodlands.

Garden Phlox, **P. paniculata.** This is the commonly cultivated perennial Phlox that imparts an abundance and variety of color to summer gardens and is often found in the wild blooming well into autumn. It is very leafy—with 15 or more nodes below the inflorescence—and except for the next species is the only one that has broad leaves with conspicuous lateral veins branching out from the midvein and joining along the margins. It is a stout, erect plant up to 6 feet high, with a large compound, pubescent panicle. Its flowers cover a wide range of colors.

P. amplifolia has even wider leaves, also with prominent side veins, but there are fewer of them than in *P. paniculata*. Its inflorescence is paniculate but glandular-hairy, and the flowers are pink-purple.

Wild Sweet William, **P. maculata.** The specific epithet refers to reddish spots or streaks on its erect stem. The inflorescence differs from our other species in being cylindrical, considerably taller than wide. The stamens are about the same length as the tube. Stream banks, meadows, and roadsides.

P. ovata, Mountain Phlox. This is a slender, decumbent plant with only about 4 nodes below the inflores-

cence. It is pubescent in the inflorescence, but the corolla tube is smooth. In this and the remaining species,
the flowers are in flat- or round-topped corymbiform
cymes and the stamens are exserted. All may be found
along roadsides and in other open areas.

P. buckleyi is a rare endemic restricted to a few dry,
rocky slopes in the northern part of our range. Its inflorescence and the corolla tube are glandular-hairy.

Carolina Phlox, **P. carolina** (plate 395), and Smooth
Phlox, **P. glaberrima,** are very similar, and together with
P. maculata they make up a taxonomically difficult group.
Both have erect, usually smooth stems with numerous
nodes below the inflorescence and narrow sharp-pointed
leaves. Typically, the leaves of *P. carolina* are relatively
wider, and its calyxes longer and more cylindrical than
campanulate, when compared with *P. glaberrima.*

Polemonium reptans, Jacob's Ladder, Greek Valerian
(plate 396). This is our only representative of the genus, which can be distinguished from *Phlox* by its pinnately compound instead of simple leaves. In the case
of *Polemonium reptans,* the principal leaves have from
11 to 21 leaflets, each 1 to 1 $^1/_2$ inches long. The attractive blue flowers are in loose panicles and are $^1/_2$ to $^3/_4$
inch long, bell-shaped with 5 lobes. In this species the
stamens do not protrude beyond the corolla. Moist
woods and clearings. Spring.

HYDROPHYLLACEAE, WATERLEAF FAMILY

Hydrophyllum virginianum, Waterleaf (plate 397).
Waterleaf flowers are in repeatedly branched cymes and
have 5-lobed tubular-campanulate corollas with 5 conspicuous, long-protruding stamens. They flower in spring
to early summer in rich woods. *H. virginianum* is smooth
or short-hairy, although the sepals have spreading hairs
on their margins. It has cauline leaves that are pinnately
divided into 5 or sometimes 7 sharply toothed leaflets
and are frequently mottled. The flowers may be either
white to pale purple or deep violet, and are $^3/_8$ inch long.

H. macrophyllum (plate 398) is densely hairy, and its
stem leaves are pinnately 7- to 13-cleft into coarsely

toothed segments. The flowers are sordid white, ¹/₂ inch long, with spreading lobes.

The cauline leaves of **H. canadense** extend above the inflorescence and are palmately 5- to 9-lobed, cordate at the base. Its flowers are white to pink-purple, ³/₈ inch long.

Phacelia fimbriata, Fringed Phacelia (plate 399). Our species of *Phacelia* grow in woods or sometimes fields or along roadsides, and have attractive 5-lobed campanulate flowers in a cyme that is coiled at first but soon elongates. The first three species sometimes put on extensive massed displays of bloom in spring. *P. fimbriata* is only a few inches high and has white flowers ³/₈ inch wide, with deeply fringed lobes.

Miami Mist, **P. purshii,** is a weak, sprawling plant with more numerous flowers similar to those of *P. fimbriata* but pale blue and less deeply fringed.

In **P. dubia** (plate 400) the corolla is also blue or whitish, but is slightly smaller and has entire lobes. Very similar to this is **P. maculata,** which sometimes occurs on rock outcrops and has bluish purple flowers. Its calyx lobes are hairy, linear, and ¹/₄ inch long, while those of *P. dubia* are narrowly ovate and shorter.

P. bipinnatifida, Purple Phacelia (plate 401), is an erect, hairy plant, glandular in the inflorescence, with long-stalked, bipinnately divided cauline leaves. Its flowers are ¹/₂ inch broad, violet-blue, with entire lobes and long-exserted stamens.

Boraginaceae, Borage Family

Plants of the Borage Family have 5-lobed flowers, regular with but a single exception. In most genera they are arranged in one-sided scorpoid racemes that uncoil as they bloom.

Echium vulgare, Viper's Bugloss, Blue Weed (plate 402), is our only species with bilaterally symmetric flowers, the upper lip being the longer. They are bright blue (but pink in bud), ¹/₂ to ³/₄ inch long, with red filaments protruding conspicuously beyond the corolla. The entire plant, including the short, tightly coiled flowering stems, is very bristly. Roadsides, fields, and waste places. Summer–fall.

The strange common name of this plant seems to beg for an explanation. The word "bugloss" (pronounced bew´gloss) translates to "ox-tongue," apparently in reference to the leaves. Echium is from the Greek word for viper, and this has been traced to a fancied similarity between the seeds and a serpent's head.

Mertensia virginica, Virginia Bluebell, Cowslip (plate 403). Another species with relatively large blue flowers is *Mertensia virginica*. This is a smooth, glaucous plant with obovate leaves 6 inches long or more. Its trumpet-shaped flowers are $^3/_4$ to 1 inch long, the tube very narrow and the limb abruptly expanded into an unlobed bell. They are pink at first, then blue, and dangle in clusters. Bottomlands. Spring.

Myosotis scorpioides, Forget-me-not (plate 404). This introduced species is the best known of the Forget-me-nots and is often cultivated. It is a loosely branched, sprawling plant with bractless racemes of $^1/_4$-inch flowers, blue with a yellow eye. The hairs on the calyxes are straight, and the lobes much shorter than the tube. Along brooks and in other wet places. Spring–fall.

Smaller Forget-me-not, **M. laxa,** is a native species but has the same habitat preferences. It is similar to the preceding but with smaller flowers (about $^1/_8$ inch wide), light blue with a yellow eye, and the calyx lobes are as long as the tube.

M. arvensis also has 1/8-inch blue flowers, but can be distinguished from the first two species by the fact that the calyx has some hooked hairs. Lawns, fields, and waste areas. Spring–summer.

M. verna, Spring Forget-me-not. This species has an erect, sometimes branched stem, up to 16 inches high, with oblong leaves under 2 inches long. The flowers are white, less than $^1/_8$ inch wide, and the calyxes are slightly irregular, with some of its hairs hooked at the end. Scorpion Grass, **M. macrosperma,** is very similar but loosely branched, with leaves up to 3 inches and calyxes with many long hooked bristles. Open woods and fields. Spring–summer.

Lithospermum canescens, Hoary Puccoon (plate 405). This showy plant is quite different from our other representatives of the Borage Family. Its flowers are bright orange-yellow, $^3/_8$ inch wide, in cymes crowded into a densely flowered, rather flat terminal inflorescence. The plant is soft-hairy and very leafy. Dry, open areas. Spring.

L. arvense, Corn Gromwell, is rough-hairy with alternate linear-lanceolate leaves $^1/_8$ to $^1/_4$ inch wide with only a midvein. The flowers are solitary in the crowded upper leaf axils, white or bluish, funnelform, $^1/_4$ inch long. Waste places. Spring–summer.

L. latifolium, American Gromwell, has ovate-lanceolate leaves that are much wider and have several distinct lateral veins. Its flowers are also axillary but are yellowish white. Dry woods and thickets. Late spring–early summer.

Onosmodium virginianum, False Gromwell (plate 406). *Onosmodium* is unusual in having flowers in which the style, which is threadlike, extends beyond the mouth of the corolla. The flowers are tubular, only slightly widened at the summit, with pointed lobes. *O. virginianum* is 1 to 2 feet high and has creamy white to yellow flowers about $^3/_8$ inch long.

O. molle and **O. hispidissimum** have yellowish or greenish white flowers. The first has fine, soft hairs; the second is a larger plant with coarse, spreading hairs. Dry woodlands. Summer.

Cynoglossum virginianum, Wild Comfrey. The flowers of this species are salverform, 3/8 inch wide, pale blue or sometimes white; they are borne in bractless racemes. The basal leaves are elliptic to ovate, up to 8 inches long; the few clasping cauline leaves are restricted to the lower half of the 2-foot stem. Open woods. Spring–early summer.

C. officinale, Hound's Tongue, is similar but leafy throughout. The lower leaves are oblanceolate and stalked, the upper ones shorter and sessile. The corolla is ¹/₄ inch wide and reddish purple. Fields, roadsides, and waste areas. Late spring–summer.

Hackelia virginiana, Stickseed, Beggar's Lice, has an open, branched inflorescence with small, if any, bracts in the upper portion of the racemes. Basal leaves are usually absent at flowering time; the cauline ones are tapered at both ends. Flowers are tiny—less than ¹/₈ inch wide—and white. Dry areas. Summer–fall.

VERBENACEAE, VERVAIN FAMILY

Verbena spp., Vervain. The native species of *Verbena* are quite different from the relatively large-flowered garden and window-box forms. Ours have small, tubular flowers with 5 spreading, slightly unequal lobes in long, slender spikes; only a few flowers in each spike open at a time. All have toothed leaves.

Verbena hastata, Blue Vervain (plate 407), may grow to 5 feet high, with several erect branched spikes with violet-blue flowers ³/₁₆ inch wide. The leaves are stalked and lanceolate, the lower ones sometimes 3-lobed at the base. Moist fields. Summer to fall.

V. stricta, Hoary Vervain, is almost as tall, with mostly solitary spikes, and is white-hairy. Its leaves are sessile and ovate. The flowers are ³/₈ inch wide, bluish purple. Open, disturbed areas. Summer.

Narrow-leaved Vervain, **V. simplex,** is less than 2 feet high and has narrow lanceolate leaves with long-tapering bases and slender solitary spikes. The flowers are ³/₁₆ inch wide, purple or lavender. Roadsides. Late spring–summer.

Our only consistently white-flowering species is **V. urticifolia** (plate 408). It grows to 5 feet high and has very slender branched, divaricate spikes with a few scattered, minute white flowers. Open woods, moist fields, and waste places. Summer–fall.

Phyla lanceolata (*Lippia lanceolata*), Fog Fruit (plate 409), is a low plant sometimes rooting at the nodes and bearing slightly irregular 4-lobed pink or lavender flowers only 1/8 inch long in dense globose heads, each of which is held above the subtending leaves on a stiff stalk. The leaves are lanceolate and toothed except near the base, up to 3 inches long. Wet, sandy areas. Summer–fall.

LAMIACEAE, MINT FAMILY

Several characteristics help us to recognize plants of the Mint Family: square stems, paired simple leaves, and two-lipped bilaterally symmetric flowers. There are exceptions, of course, and these same features also appear in other families. Many have aromatic foliage, among them lavender, thyme, sage, catnip, and of course various "mints."

> Flowers in dense headlike clusters: *Monarda, Blephilia,*
> *Satureja* (part), *Pycanthemum, Origanum*
> Flowers in dense spikes: *Teucrium, Agastache, Physostegia,*
> *Nepeta, Mentha* (part), *Prunella*
> Flowers in dense axillary whorls: *Galeopsis, Leonurus, Mentha*
> (part), *Lycopus, Marrubium, Hedeoma, Lamium, Glechoma*
> Flowers in loose or few-flowered inflorescences: *Collinsonia,*
> *Trichostema, Cunila, Satureja* (part), *Stachys, Meehania,*
> *Synandra, Salvia, Perilla, Melissa, Scutellaria*

Flowers in dense headlike clusters

Monarda punctata, Horse Mint. The *Monardas* have flowers with narrow, markedly 2-lipped corollas, radiating from dense rounded heads which are subtended by showy bracteal leaves. Horse Mint has axillary as well as terminal flower clusters. The individual flowers are $^3/_4$ inch long, creamy white or yellowish, dotted with purple; the 2 stamens are included beneath the strongly arched upper lip. The bracteal leaves may be partially or wholly lavender or whitish. Dry soil. Summer–fall.

The name Horse Mint is sometimes applied to all Monardas *as a group, the word "horse" being intended merely to characterize a plant as comparatively coarse or large (note also Horse Balm in this family). Names of other animals such as dog, cow or hog are often used as well, frequently in a derogatory sense.*

M. fistulosa, Wild Bergamot (plate 410). This and the remaining species of *Monarda* share several features: The flower clusters are solitary and terminal on the branches. The upper lip of the corolla is straight and slightly exceeded by the stamens, and the bracteal leaves are often tinged with the basic color of the flowers. M. *fistulosa* has 1-inch pink or lavender flowers. The upper lip bears a small but distinct tuft of white hairs at its apex, a unique characteristic that sets it apart from our other species. Open woods and meadows. Summer–fall.

M. clinopodia, Basil Balm, is similar but has a whitish corolla with purple spots. Woods. Late spring–summer.

M. didyma, Oswego Tea, Bee Balm (plate 411). The flowers of Oswego Tea have long been a favorite of gardeners. They are 1 to 1 $^1/_2$ inches long and typically bright crimson. Many color variants occur, including a purple-flowered form sometimes referred to as **M. media,** and thought by some to be a hybrid with M. *fistulosa* or M. *clinopodia*. Moist woods and stream banks. Summer–fall.

Blephilia ciliata, Downy Wood Mint. In this genus the rounded, headlike flower clusters—or at least the uppermost ones—are crowded together on the stem. As in *Monarda*, there are only 2 stamens, but the calyx is irregularly 2-lipped instead of having 5 equal teeth. B. *ciliata* has entire or obscurely toothed leaves, sessile or very short-stalked. The flowers are $^1/_2$ inch long, pale lavender with purple spots. Dry woods. Late spring–summer.

B. hirsuta, Hairy Wood Mint, is similar but its leaves are sharply toothed and the petioles are $^3/_8$ inch long or more. Moist woods. Summer.

Satureja vulgaris, Wild Basil (plate 412), has a compact, rounded terminal flower cluster, sometimes with 1 or 2 in the upper leaf axils as well. The flowers are about $^1/_2$ inch long, purplish pink, with 4 stamens, and are intermixed with long, bristly bracts. This is a hairy plant with leaves up to 1 $^1/_2$ inches long, entire or with a few low teeth. Open woods and roadsides. Summer–fall.

Very different is **S. calamintha var. nepeta,** which has smaller leaves and many well-spaced axillary clusters of 5 to 10 flowers each, with $^3/_8$-inch corollas. Fields and waste ground.

Pycnanthemum spp., Mountain Mint. These are freely branched plants with a strong minty taste, growing in dry to moist wooded or open areas and blooming from summer to fall. The flowers are small, white to pinkish or purplish and usually spotted with purple; there are 4 stamens. The inflorescences consist of dense rounded heads, both terminal and axillary.

Pycnanthemum incanum, Hoary Mountain Mint (plate 413). The upper portions of this plant—most conspicuously the bracteal leaves—appear white-powdery from a fine pubescence. Its principal leaves are ovate-lanceolate, sharply toothed, and whitened beneath, on petioles up to $^1/_2$ inch long. There are both terminal flower heads and secondary ones in the axils, $^1/_2$ to more than 1 inch in diameter.

In **P. montanum** the leaves are lanceolate, from $^1/_2$ to 1 $^1/_4$ inch wide and tapered at the base to stalks about $^1/_4$ inch long, and have sharply toothed margins. The flower heads are $^1/_2$ to $^3/_4$ inch across, both terminal and sessile in the axils of the upper leaves.

P. muticum, Short-toothed Mountain Mint, is gray-hairy above, and has ovate-lanceolate leaves rounded at the base, with a few low teeth, on petioles less than $^1/_8$ inch long. It has small flower heads ($^1/_4$ to $^1/_2$ inch wide), mostly terminal.

P. tenuifolium, Narrow-leaved Mountain Mint (plate 414). This and the remaining species are much-branched, very leafy, and floriferous. True to its name, this smooth-stemmed species has entire linear leaves only about $^1/_8$ inch wide. The flowers are borne in clusters no more than $^3/_8$ inch across, which in turn make up a more or less flat-topped inflorescence.

P. virginianum, Virginia Mountain Mint, is similar but has entire linear-lanceolate leaves up to $^3/_8$ inch wide, rounded at the base. Its stem is pubescent on the angles.

P. verticillatum, Torrey's Mountain Mint, has hairy stems and lanceolate leaves up to $^1/_2$ inch wide, tapered at the base, with a few low teeth on the margins.

Origanum vulgare, Wild Marjoram, a rare escape, is an erect, hairy plant with entire ovate leaves, which may be yellow. Its inflorescence is a trichotomously branched panicle of pink to magenta flowers in crowded cymes with purplish bracts. The lower lip is deflexed and deeply 3-lobed, and the lower stamens are exserted and divergent. Summer–fall.

Flowers in dense spikes

Teucrium canadense, American Germander, Wood Sage (plate 415). This is a tall plant with numerous $5/8$-inch, pinkish purple flowers crowded into a long, dense, spikelike raceme. The 4 stamens are arched upward; the upper corolla lip is apparently absent, and the lower one has 2 pairs of small lateral lobes and a broad scoop-shaped terminal lobe. Moist woods, thickets, and meadows. Summer.

Agastache scrophulariaefolia, Purple Giant Hyssop (plate 416). A stout, hairy plant 5 feet high or even taller. The purplish flowers are $1/2$ inch long, with 4 protruding stamens. They are crowded into cylindrical terminal spikes. **A. nepetoides,** Yellow Giant Hyssop, is essentially smooth and has greenish yellow flowers. Open woods and thickets. Summer.

Physostegia virginiana *(Dracocephalum virginianum),* Obedient Plant, False Dragonhead (plate 417). This plant is frequently cultivated, not only for the beauty of its flowers but because of their curious habit of "obediently" remaining in whatever position they have been bent. They are 1 inch or longer, rose-pink with purple veins, and are closely and evenly spaced along the axis of a long spike, each in the axil of a small bract. The leaves are narrowly lanceolate and sharply toothed. Moist woods and thickets. Summer–fall.

Nepeta cataria, Catnip. This abundant weed is downy throughout. Its leaves are triangular, cordate-based, and coarsely toothed, on distinct petioles. The flowers are $3/8$ inch long, white with pink or purple dots, the 4 stamens ascending beneath and concealed by the upper lip of the corolla. They are in short spikes of crowded clusters terminating the branches. Waste areas. Summer–fall.

Mentha spp. These are the true mints. They are aromatic and grow in wet ground, blooming in summer, sometimes into the fall months. The corollas are 4-lobed and almost radially symmetric, the upper lobe often notched, and there are 4 protruding stamens. Our first three species have their flowers in terminal spikes made up of clusters each of which is subtended by small bracts; in the other two the clusters are more widely separated and subtended by foliage leaves.

Mentha piperita, Peppermint (plate 418), has thick, compact spikes of pale pink or purplish flowers and smooth, lanceolate leaves with petioles from $^1/_8$ to $^1/_4$ inch long.

Many uses have been found for the cooling taste and fragrance of Peppermint, but one of its most unorthodox applications originated with Native American hunters who steeped their steel traps in a strong, pungent infusion of the leaves in order to mask the telltale odor of blood before resetting them.

M. spicata, Spearmint, can easily be told from Peppermint by its distinctive aroma. Also, its spikes are more slender and somewhat interrupted, and its leaves, also smooth and lanceolate, are virtually sessile.

M. rotundifolia, Round-leaved Mint, is characterized by its round-ovate, scalloped leaves, which are densely soft-hairy and prominently veined on both sides. The spikes are continuous for most of their length, and the flowers are white with purple spots. Escaped to roadsides and waste ground. Late spring to fall.

M. cardiaca, Small-leaved Mint, has flowers that are in dense whorls in the upper axils and are subtended by leaves that are smaller than the ordinary foliage leaves but extend beyond the flower clusters.

M. arvensis, Field or Wild Mint, is our only native species of *Mentha*. The bracteal leaves are as large as the stem leaves and greatly exceed the flower clusters.

Prunella vulgaris, Self Heal (plate 419). A very widespread species, Self Heal apparently has been able to adapt to a great variety of habitats. It is a rather small plant with short, dense cylindrical spikes of violet flowers $^3/_8$ to $^5/_8$ inch long. The upper lip is hoodlike, arching over the 4 stamens, the lower lip bent downward and its median lobe fringed. Spring–fall.

Although Self Heal is almost ubiquitous, it never seems to wear out its welcome by over-competing with other vegetation. Moreover, its flowers are attractive enough to compare favorably with some of our small terrestrial orchids. Its other name, Heal All, presumably exaggerates its medicinal value, for the record does not indicate that it ever approached the status of a panacea.

Flowers in dense axillary whorls

(See also Mentha)

Galeopsis tetrahit, Hemp Nettle (plate 420), has a bristly stem swollen at the nodes and hairy, well-stalked leaves. The flowers are white to pink, $^1/_2$ to $^3/_4$ inch long, the upper lip bristly, the lower 3-lobed, marked with yellow and purple, and bearing a pair of basal protuberances; there are 4 stamens. Fields, roadsides, and waste places. Summer.

Leonurus cardiaca, Motherwort (plate 421), is an erect plant up to 5 feet tall. Its leaves are palmately lobed and sharply toothed, becoming progressively smaller above, the upper ones unlobed but with 3 large teeth; all extend more or less horizontally from the stem. The flowers are pink, $^3/_8$ inch long, with a white-hairy upper lip and 4 stamens, crowded into the leaf axils. The calyx lobes end in stiff spines. Waste ground and roadsides. Summer.

Lycopus spp. The plants in this genus have tiny white 4-lobed flowers, almost regular, the uppermost lobe slightly larger; there are 2 stamens. They are packed into small clusters in the axils of widely spaced pairs of leaves. Wet places, flowering in summer.

Lycopus virginicus, Bugleweed (plate 422). The foliage of *Lycopus virginicus* varies from dark green to deep purple. The leaves are ovate with a long-tapering, petiole-like base, and are coarsely toothed. **L. uniflorus** (*L. sherardii*) is similar, but the leaves are light green and lanceolate.

L. americanus, Water Horehound. In this species the leaves are lanceolate, most of them pinnatifid with blunt lobes.

Marrubium vulgare, Horehound (plate 423). This is a downy plant with deeply-veined, blunt-toothed ovate leaves 1 to 2 inches long. The flowers are white, ¼ inch long, with 4 stamens; the calyx is regular and has 10 teeth. Waste ground. Late spring–fall.

Hedeoma pulegioides, American Pennyroyal, is a strongly aromatic, usually branched plant with numerous pairs of lanceolate or elliptic leaves 1 inch long or less, entire or with small teeth. Each axillary cluster bears bluish flowers only ⅛ inch long and just slightly surpassing the calyx, with 2 stamens. Dry fields and open woods. Summer–fall.

This indigenous herb was utilized first by Native Americans and then by the European settlers as a stimulant, to reduce fever by inducing perspiration, to relieve headaches, and for delayed menstruation. Rubbing the crushed plant on the skin was a pleasant way to ward off chiggers and other insects—much nicer than most other contemporary preparations, such as turpentine, which tended to be equally repellent to the user.

Lamium spp. Our species of *Lamium* are common weeds of lawns, fields, and waste places, flowering from spring until fall. They are weak-stemmed but erect plants. Their leaves, which are about the same size throughout, have cordate bases and deeply scalloped margins. There are 4 stamens.

Lamium amplexicaule, Henbit (plate 424). In Henbit the leaves are nearly round and the upper ones sessile, sometimes clasping. The magenta flowers are about ½ inch long and almost erect. The upper lip is conspicuously dilated at the throat.

L. purpureum, Dead Nettle, has stalked ovate leaves, most concentrated in the upper portion where they are strongly tinged with purple. Its reddish purple flowers are less erect than those of Henbit.

Glechoma hederacea, known as Ground Ivy or Gill-over-the-Ground (plate 425), is a creeping herb often forming mats in disturbed areas. Its leaves are round or kidney-shaped, petiolate, with blunt teeth. Its ¹/₂-inch flowers are violet, striped and spotted with red within, and have 4 stamens. Spring–summer.

Flowers in loose or few-flowered inflorescences
(See also Satureja)

Collinsonia canadensis, Horse Balm (plate 426), has an open pyramidal panicle of lemon-scented flowers ¹/₂ inch long, light yellow, the lower lip extended and fringed. The 2 stamens and pistil are long-exserted. This is a large, coarse plant with several pairs of ovate leaves up to 8 inches long. Rich woods. Summer–fall.

C. verticillata is shorter and usually has only 4 leaves crowded together on the stem. Its flowers are purplish and have 4 stamens, and the inflorescence is usually unbranched. Woods. Spring–summer.

Trichostema dichotomum, Blue Curls (plate 427). This is a slender, stiffly branched plant with entire oblong to lanceolate leaves and 1 to 3 slightly irregular 5-lobed blue flowers on axillary peduncles. They are ¹/₂ to ³/₄ inch long; the 4 blue stamens extend far beyond the corolla and are strongly curved, hence the common name. The calyx is strongly 2-lipped. **T. setaceum** is similar but has linear leaves less than ¹/₈ inch wide. Dry soil. Summer–fall.

T. brachiatum (*Isanthus brachiatus*), False Pennyroyal, has small (under ¹/₄-inch) light blue flowers with almost radially symmetric corollas and calyxes. Its stamens are straight and only slightly exserted. Dry, basic soil. Early fall.

Cunila origanoides, known as Dittany or Stone Mint, is a wiry, much-branched plant with sessile, ovate leaves. Its 5-lobed flowers are ³/₈ inch long, purplish to white, with 2 protruding stamens; the corolla is almost radially symmetric. They are borne in small terminal and axillary clusters. Dry or rocky woods and clearings. Summer–fall.

Stachys spp., Hedge Nettle. The Hedge Nettles are variable and probably hybridize and therefore are difficult to identify. They are tall plants with interrupted spikes of $^1/_2$-inch pink to purple flowers, which are 5-lobed; the upper part ends in an inverted cup that covers the 4 stamens, and the lower is bent sharply downward.

Stachys latidens (plate 428) is our most common species. Its stems are mostly glabrous on the sides but have short, reflexed hairs on the angles. Woods and clearings. Summer.

S. clingmanii differs from the preceding in having long, spreading bristles on the angles of its stems. It is found mostly in high altitude forests.

In **S. nuttallii** (*S. riddellii*), the stems are not only villous but also bear very small glandular hairs. Moist woods and low meadows.

Meehania cordata (plate 429) is a trailing plant with ascending flowering stems up to 8 inches high. Its leaves are heart-shaped and petioled. The few attractive blue to lavender flowers are large (1 inch or more long) and are crowded into one-sided terminal spikes. Rich woods. Late spring–early summer.

Synandra hispidula. Large, striking flowers and comparative rarity combine to make *Synandra* a very worthwhile find. Its ovate leaves are cordate at the base and, except for the uppermost pair, are long-stalked. There are several flowers in an interrupted terminal spike; they are about 1 $^1/_4$ inches long, white with dainty lavender stripes on the lower lip. Rich woods. Spring–early summer.

Salvia lyrata, Lyre-leaved Sage (plate 430). The principal leaves of this plant are in a basal rosette and are pinnately but irregularly lobed, the terminal segment broadly rounded. Its flowers are about 1 inch long, blue-violet, growing in few-flowered whorls in an interrupted spike. There are only 2 stamens. Meadows and disturbed areas. Spring–summer.

In **S. urticifolia,** Nettle-leaved Sage, there are no basal leaves; the cauline leaves are ovate and sharp-pointed at the apex, with toothed margins. Its flowers are blue and seldom more than $^1/_2$ inch long. Dry woods. Spring–summer.

Perilla frutescens, Beefsteak Plant. Sometimes cultivated for its foliage, which may be reddish purple, Beefsteak Plant has escaped to roadsides and dry woods where it blooms during summer and fall. Its leaves are large (up to 6 inches long), ovate, tapered to a long point, and long-stalked. The flowers are less than $^1/_4$ inch long, purple or white, one of the 5 lobes slightly longer than the others; there are 4 stamens. They are borne in loose, elongate terminal and axillary spikelike racemes.

Melissa officinalis̄, Lemon Balm. This is another cultivated plant that has escaped to disturbed areas. Its leaves, which are lemon-scented, are ovate with scalloped margins and have few-flowered clusters in their axils. The corollas are pale blue and about $^1/_2$ inch long, with very short lips, and curve upward. Summer–fall.

Scutellaria spp., Skullcap. The field mark of Skullcaps, for which they are named, is a small but conspicuous protuberance on the upper lip of the calyx. The flowers are mostly blue or violet, curved upward, with a helmet-shaped upper lip covering the 4 stamens. In our species they are borne in terminal and lateral racemes.

> Flowers $^3/_4$ inch or more long: *S. integrifolia, S. serrata, S. incana*
> Flowers less than $^3/_4$ inch long: *S. elliptica, S. saxatilis, S. ovata, S. lateriflora*

Scutellaria integrifolia, Hyssop Skullcap (plate 431). Unlike our other species, the middle and upper leaves of Hyssop Skullcap have entire margins; they are lanceolate and taper to a sessile base. The lower leaves (which do not always persist until flowering time) may be ovate, toothed, and petiolate. The flowers are $^3/_4$ to 1 inch long in one or a few terminal racemes. Open woods and clearings. Spring–summer.

S. serrata, Showy Skullcap. With flowers 1 inch or more in length, this species is well named. It is a smooth plant with 2- to 4-inch ovate, toothed leaves, their bases narrowed abruptly to long petioles. The raceme is solitary, up to 6 inches long. Woods. Spring–early summer.

S. incana var. **punctata,** Downy Skullcap, is a somewhat hoary plant with stalked, scalloped, ovate-lanceolate leaves that taper to a petiole. The $^3/_4$- to 1-inch flowers are borne in numerous racemes. Woods and clearings. Summer–fall.

S. elliptica, Hairy Skullcap. The pubescence on the stems of this species may consist of spreading or curving hairs, sometimes mixed with glandular hairs. Its leaves are ovate, tending to be rhombic, with scalloped margins and blunt at the apex. There are a few racemes of small ($^1/_2$- to $^3/_4$-inch flowers). Woods. Spring–summer.

S. saxatilis, Rock Skullcap, is a weak-stemmed plant with triangular or heart-shaped leaves less than 2 inches long, with rounded teeth on the margins. The $^5/_8$-inch flowers are in 1 to 3 racemes which often are one-sided. Rocky woods. Late spring–summer.

S. ovata, Heart-leaved Skullcap. This is an extremely variable species but can be identified by its leaves, which have heart-shaped bases and blunt-toothed blades up to 4 inches on long stalks. The flowers are between $^1/_2$ and $^3/_4$ inch long. Woodlands. Summer.

S. lateriflora, Mad-dog Skullcap, has the smallest flowers among our species—only about $^1/_4$ inch long. The racemes are numerous in the leaf axils and are one-sided. Moist or wet woods and thickets. Summer–fall.

Mad-dog Skullcap was among the earliest plants to be brought back to Europe from America in the 1600s. It was of particular interest to herbalists because of its reputation as a supposed remedy for rabies.

Solanaceae, Nightshade Family

Physalis spp., Ground Cherry. Our species have broadly bell-shaped, yellow flowers, the segments very weakly 5-lobed, usually with 5 brown or purplish blotches in the throat. The fruit is a berry enclosed by the calyx, which enlarges to become a parchmentlike husk. Leaves often have unequal bases; intraspecific variations in shape and dentition make positive identification difficult. Ground Cherries will be found in a variety of open, often disturbed, habitats. Flowering times range from spring until fall.

Physalis heterophylla, Clammy Ground Cherry (plate 432), is invested with short hairs, some sticky-glandular. Its leaves are ovate, rounded at the base, with irregular large teeth. The flowers are $1/2$ to $3/4$ inch long, with yellow anthers.

P. virginiana has soft reflexed hairs and ovate to lanceolate leaves tapered at both ends, with a few blunt teeth. The $5/8$-inch corolla has yellow or bluish anthers. Similar but essentially smooth and with entire or wavy-margined ovate leaves and blue anthers is **P. longifolia var. subglabrata.**

Two species with soft spreading hairs, uneven leaf bases, and $3/8$-inch flowers with blue anthers are **P. pubescens** and **P. pruinosa.** In the first, the leaves have cordate bases and margins that are entire or shallowly toothed only near the summit; in the second, they are not indented at the base, are unevenly toothed throughout, and are covered by a dense gray down that obscures their green color.

P. angulata is a smooth plant with small flowers that lack the usual dark spots; the anthers are blue. Its leaves are ovate with coarse irregular teeth.

Nicandra physalodes, Apple-of-Peru. As the specific epithet would suggest, there are similarities between this plant and some in the genus *Physalis*: for example, the broadly campanulate, shallowly lobed corolla—although its color (light blue) and large size (1 to 1 $1/2$ inches wide) are exceptions—and the berry enclosed by an inflated calyx. Its leaves are up to 8 inches long and are coarsely and very unevenly toothed. Fields and waste places. Summer–fall.

Datura stramonium, Jimson Weed, Thorn Apple (plate 433). This is a coarse, branched plant up to 5 feet high. The leaves are ovate, with large, coarse angular teeth. The flowers are trumpet-shaped, each of the 5 flaring lobes ending in a long, sharp tooth;

they are up to 4 inches long, white or lavender-tinged. The calyx is 5-sided, narrowly winged, and half as long as the corolla. Fields, roadsides, and waste places. Summer–fall.

This attractive but dangerous weed, often seen in barnyards, pastures, and vacant lots, has been responsible for many fatal poisonings, some due to the brewing of a "tea" from the leaves in the mistaken belief that it would relieve asthma. The most famous episode involved the deaths of an entire detachment of soldiers who were sent to Jamestown, Virginia, in 1676 and sought to experience hallucinogenic effects from a similar concoction; this incident gave the plant the name "Jimson Weed" (a corruption of "Jamestown").

Solanum carolinense, Horse Nettle (plate 434). All of our species of *Solanum* can be found along roadsides and in fields and waste places, blooming from late spring until fall. One feature that they all share is that the anthers are connivent, forming a beak around the style, and open by terminal pores. Horse Nettle has scattered but sharp prickles, and leaves with shallow lobes. The star-shaped corollas are 1 inch across, white, light blue, or lavender. Its berries are yellow.

S. rostratum, Buffalo Bur (plate 435), is an immigrant from the West. It is much more thickly armed and is our only species in which the fruit is completely enclosed by the calyx—which also is spiny. The flowers are yellow, 1 inch in diameter, and its leaves are deeply pinnately lobed.

Black Nightshade, **S. ptycanthum** (*S. nigrum, S. americanum*), is a thornless plant and has triangular-ovate leaves which may have rounded teeth. The corolla has 5 reflexed segments and is only about ¼ inch across. The flowers are white; it is the berries that are black.

Bittersweet Nightshade, **S. dulcamara** (plate 436), is an unarmed twining vine with ovate leaves up to 4 inches long, often with 1 or 2 small diverging basal lobes. Its corolla is bright violet with recurved segments, ½ inch or more across. The berries are crimson.

SCROPHULARIACEAE, FIGWORT FAMILY

Verbascum thapsus, Woolly Mullein, Flannel Plant, Aaron's Rod (plate 437). Although this family is known for having bilaterally symmetric flowers, those of the Mulleins come close to having 5 identical corolla lobes, the lower ones being only slightly larger. Woolly Mullein is a tall, rigidly straight plant, seldom branched, with crowded alternate leaves densely covered with springy branched hairs, their bases decurrent on the stem. The upper part of the plant consists of a continuous spike about $^5/_8$ inch in diameter, packed with yellow flowers that open randomly. A familiar weed of roadsides, fields, and waste places. Summer.

Mullein was known in ancient Rome, where the large dried seed stalks were impregnated with wax and used as torches—a practice modified centuries later by the Eastern Cherokees, who substituted animal fat.

V. phlomoides is rather similar but often branched and the inflorescence interrupted; its flowers are larger than those of *V. thapsus*. The leaves are not decurrent but merely clasp the stem. Fields. Late spring–early summer.

Moth Mullein, **V. blattaria** (plate 438), is a more slender plant with a few branches and smooth, toothed, or scalloped leaves. The flowers are in an open raceme, 1 to 1 $^1/_4$ inch across, either yellow or white, with purple hairs on the filaments of the stamens. Fields and roadsides. Late spring–early summer.

Aureolaria spp., False Foxglove. Our species of *Aureolaria* (sometimes placed in the genus *Gerardia*) bloom from summer to fall in open woods, where they are thought to be semiparasitic on oak roots. The flowers are bright yellow, between 1 and 2 inches long, with a bell-shaped tube and 5 almost equal spreading lobes, on pedicels that sometimes curve upward. The leaves are opposite.

Aureolaria virginica (plate 439) has a downy stem. The lower leaves have a few large lobes near the base; the upper tend to be entire.

A. flava is one of two smooth species, its stem glaucous and often purplish. The lower leaves are pinnately cleft. In **A. laevigata** the stem is green, not glaucous; most of its leaves are entire, but some lower ones may be shallowly lobed or toothed.

A. pedicularia (incl. **A. pectinata**) is widely branched, and is characterized by its pinnately incised, fernlike foliage, as well as by its sticky glandular hairs.

Agalinis purpurea, Gerardia (plate 440). Like *Aureolaria*, plants in this genus may also be found listed under the scientific name *Gerardia*. They are all wiry plants with linear leaves and racemes of pink flowers that are only weakly bisymmetric, blooming in summer and fall. A. *purpurea* has leaves up to 1 ½ inches long and ⅛ inch wide, and flowers between ¾ and 1 ½ inches long on extremely short stalks. Moist woods and meadows.

A. setacea has threadlike leaves up to 1 ¼ inches long. Its flowers are between ⅝ and 1 inch long, on ¾-inch pedicels. All of the corolla lobes are spreading; the throat is open and woolly near the base within. Dry, sandy soil.

A. tenuifolia is similar but has wider leaves and flowers that are only ½ inch long. The upper lip is arched over the stamens, nearly closing the throat, which is smooth within. Dry woods and fields.

A. decemloba has short, narrow leaves. The flowers and their stalks are each about ½ inch long. Each corolla lobe has a shallow notch in the margin, which suggested the specific epithet. Dry habitats.

Penstemon spp., Beardtongue. The flowers of *Penstemon* are bilaterally symmetric with 5 lobes, 2 above and 3 below, and are arranged in a terminal inflorescence subtended by more or less reduced opposite lanceolate leaves. There are 4 fertile stamens and a sterile one which is hairy (the "beardtongue"). They are found in fields and woodland margins and along roadsides, flowering in spring and summer.

Penstemon digitalis, known as Foxglove Beardtongue (plate 441), is a smooth plant and has 1-inch flowers, white, often with purple lines within. In this and the next two species the corolla tube is abruptly dilated to form a wide, open throat, and the lobes of the lower lip do not appreciably project beyond those of the upper.

P. laevigatus is also essentially smooth, but has smaller ($^3/_4$-inch) flowers that are pale violet outside and white inside. **P. calycosus** is very similar but finely hairy, and bears flowers up to 1 $^1/_4$ inches long.

P. smallii, Small's Penstemon (plate 442). In this and the succeeding species the corolla is gradually expanded and not strongly differentiated into a tube and throat, and the lower lobes extend beyond the upper ones. *P. smallii* has a broad, well-inflated corolla up to 1 $^1/_4$ inches long, purple but lighter at the base with purple and white lines within. The plant is minutely downy. The leaves just below the inflorescence are only slightly reduced in size, and the bracts beneath the flowers are leaflike.

P. canescens, Gray Beardtongue (plate 443), has a smaller corolla, light purple to pinkish, lined within. The plant is uniformly covered with short, grayish hairs, glandular in the inflorescence; the bracts are much smaller than the leaves. The sterile stamen is yellow-hairy and does not extend beyond the corolla tube.

In **P. australis** the flowers are reddish purple, 1 inch long, and less inflated—not more than $^1/_4$ inch in diameter. The sterile stamen is conspicuously exserted.

P. pallidus has narrow white flowers, $^3/_4$ inch long, with purple lines within, the staminode not exserted.

P. hirsutus differs from the others in that the corolla, which is 1 inch long, pale violet, and very slender, has a lower lip that arches upward, nearly closing the throat.

Chelone lyonii, Turtlehead (plate 444). We have four confusingly similar species of Turtlehead, all preferring moist or wet habitats and flowering in summer and fall. C. *lyonii* is the one most often found at high elevations. Its opposite leaves are widest below the middle and rounded at the base, and most are on petioles at least $^1/_2$ inch long. The flowers, as in all but one of our other species, are rose-purple and up to 1 ¹/₂ inches long. There are 4 fertile stamens with woolly anthers and a hairless, sterile filament that is white or rose-tipped. The inflorescence is a crowded terminal spike.

C. *obliqua* is similar but its leaves are widest near the middle and have acute bases, and the petioles are ¹/₂ inch or less. The staminode is white.

C. *glabra* (plate 445) has white flowers, although they are often strongly tinged with purple; the staminode is green. Its leaves resemble those of C. *obliqua* but are proportionately narrower, and the petioles are under ¹/₂ inch.

C. *cuthbertii* has sessile lanceolate leaves and flowers that are strongly 4-ranked. Its staminode is purple.

The rules of botanical nomenclature do not permit assigning the same generic name to two different groups of plants, but obviously such restrictions do not cross over into the animal kingdom, or we would not be using the word "Chelone" to apply equally to a genus of turtles and a genus of flowers.

Mimulus ringens, Monkey Flower (plate 446). These are erect plants with opposite lanceolate leaves, growing in wet places and flowering in summer and early fall. The attractive, light blue-violet axillary flowers are 1 inch long with a flared, ruffled lower lip and a ridged palate nearly closing the throat. In M. *ringens* the flowers are on 1- to 1 ¹/₂-inch stalks, and the leaves are sessile. The stem is square but without wings.

M. *alatus* resembles the preceding, but the flowers are on shorter pedicels and are exceeded by the petiolate leaves. The stem is narrowly winged.

Pedicularis canadensis, Wood Betony, Lousewort (plate 447). Wood Betony is a hairy plant up to 1 foot high with alternate stalked leaves that are pinnately lobed almost to the midrib. The flowers are about 1 inch long, usually bicolored yellow with purplish or brownish red, the upper lip 2-toothed at the apex and arching over the lower. They are tightly crowded into a compact terminal spike. Woods or clearings. Spring.

Swamp Lousewort, **P. lanceolata,** is taller, essentially smooth, and has mostly opposite leaves that are less deeply lobed than in *P. canadensis.* The flowers are pale yellow, ³/₄ inch long, and the upper lip is entire. Wet woods, meadows, and bogs. Fall.

These plants owe their scientific name to pediculus, *the Latin word for louse—not to indicate any possible value as a repellent, as might be imagined, but rather to recall an old European superstitious belief that cattle grazing on it would become heavily infested.*

Scrophularia marilandica, Figwort, Carpenter's Square (plate 448). At 8 feet the Figworts are among the largest of our herbaceous plants in the Scrophulariaceae. They have toothed ovate leaves 10 inches long and a loose branched inflorescence up to a foot high. The individual flowers, on the other hand, are less than ³/₈ inch in this species. They are reddish brown and strongly bilabiate, the 2-lobed upper lip erect, the median lobe of the lower lip directed downward. There is a sterile stamen, which in *S. marilandica* is purple or brown. Summer–fall.

Hare Figwort, **S. lanceolata,** is very similar except that the flowers are slightly larger and the sterile stamen is yellowish green. Also, the panicle is less widely spreading, tending to be more cylindrical than pyramidal. It blooms in spring and early summer. Both grow in open woodlands and disturbed areas.

Veronicastrum virginicum, Culver's Root (plate 449). The tallest plants in this family have some of the smallest flowers, and Culver's Root is no exception. It grows up to 7 feet high and has whorls of toothed lanceolate leaves up to 6 inches long. Its flowers are crowded in a tall terminal raceme and somewhat shorter racemes that arise from the uppermost whorl. They are white, ¹/₄ inch long or less, tubular with 4 almost equal lobes, and have 2 stamens and a style all of which protrude far beyond the corolla. Moist meadows, thickets, and stream banks. Summer.

Castilleja coccinea, Indian Paintbrush, Painted Cup (plate 450). This is our only example of a genus that is represented by a score of species in the western United

States. The tubular corollas are greenish yellow and bilaterally symmetric, the lower lip very short and deflexed. The flowers are almost hidden by 3-lobed bracteal leaves, the upper portions of which are brilliantly colored, scarlet in most plants but occasionally yellow. It is believed to be a root-parasite. Moist fields and margins of woods. Spring–early summer.

Linaria vulgaris, Common Toadflax, Butter and Eggs (plate 451). This species has 1-inch-long yellow flowers with a downward-pointing spur and an orange "palate" that nearly closes the throat of the corolla. They are crowded in a spike above the numerous linear leaves. It blooms in spring, earlier than the next species; both have become weeds along roads and in waste places.

Old-field Toadflax, **Nuttalanthus canadensis** (*Linaria canadensis*) (plate 452), is a slender plant with scattered short (1-inch) linear leaves below the middle and an open terminal raceme of flowers. These are usually less than $^1/_2$ inch long including the slender spur, and are blue with a white palate.

Chaenorrhinum minus, Dwarf Snapdragon, is one of those European immigrant plants that seem to thrive only along railway beds, where it blooms in summer. It has alternate, blunt, linear leaves and small flowers borne singly on slender stalks in the axils. The corolla, which is only about $^1/_4$ inch long including the basal spur, is strongly bilabiate and is bluish purple except for a white lower lip and the yellowish palate, which does not close the throat.

Melampyrum lineare, Cow Wheat (plate 453), is a small, branched parasitic plant of dry woodlands. Its leaves are opposite and lanceolate (some with a few large teeth near the base). The flowers are single in the upper axils; they are $^1/_2$ inch long, white with a yellow palate, fading to rose. Open woods. Spring–early summer.

Veronica spp., Speedwell. The Speedwells are among our better-known weeds, but few of us are aware of the variety that exists among the several species and their little blue or white flowers. They are 4-lobed, but are bilaterally symmetric, the upper lobe being wider than the lateral ones and the bottom lobe narrower. The principal leaves are paired.

> Flowers in racemes: *V. officinalis, V. chamaedrys, V. americana, V. anagallis-aquatica, V. scutellata, V. serpyllifolia*
> Flowers solitary: *V. persica, V. hederaefolia, V. arvensis, V. peregrina*

Veronica officinalis, Common Speedwell (plate 454), is a hairy, reclining, mat-forming plant with thick, toothed, elliptic leaves up to 2 inches long. The flowers are pale blue-violet marked with darker radiating lines, less than $^1/_4$ inch across, in spikelike racemes; the pedicels are extremely short and the subtending bracts equaling the flowers. Woodlands, fields, and roadsides.

In **V. chamaedrys** the flowers are racemose, on $^1/_8$-inch pedicels and surpassing the bracts; they are $^1/_4$ inch wide, blue, lighter in the center, with darker lines. Its leaves are ovate.

American Brooklime, **V. americana.** This species and the next two are found in wet habitats, and have their flowers in axillary racemes. *V. americana* is succulent and has toothed, lanceolate-ovate leaves on short petioles. Its flowers are pale, almost white, and only $^3/_{16}$ inch wide. Spring–summer.

Another smooth plant is Brook Pimpernel or Water Speedwell, **V. anagallis-aquatica.** Its leaves are sessile or clasping, with shallow teeth or none. The flowers are pale lavender, $^3/_{16}$ inch wide. Spring–fall.

Marsh Speedwell, **V. scutellata,** is virtually hairless, with sessile, linear leaves up to 3 inches long. Its flowers are pale violet, $^3/_{16}$ inch wide, on long, threadlike pedicels. Spring–fall.

In Thyme-leaved Speedwell, **V. serpyllifolia,** the lower leaves are $^3/_8$ inch and oval; they are replaced in the inflorescence by small bracts, making it technically a terminal raceme. The flowers are $^1/_8$ to $^1/_4$ inch wide, white or pale blue with darker lines. Fields, lawns, and open woods. Spring–summer.

V. persica, Bird's-eye Speedwell (plate 455). In the remaining species the flowers are solitary in the upper leaf axils. *V. persica* has the largest flowers (nearly ¹/₂ inch), bright blue with darker stripes and a white center, on long pedicels. Its leaves are about ¹/₂ inch long, ovate, and have a few coarse teeth. Lawns and waste areas. Spring–summer.

Ivy-leaved Speedwell, **V. hederaefolia,** has leaves about the same size but wider than long and with 3 to 5 shallow lobes near the base. The flowers are blue and less than ¹/₄ inch wide. Lawns, fields, and waste places. Spring.

In the next two species the leaves are gradually reduced in size as they progress into the inflorescence, and the flowers are nearly sessile. Corn Speedwell, **V. arvensis,** is hairy, and the foliage leaves are ¹/₂ inch long, ovate with a few teeth, and palmately veined. Its bright blue flowers are less than ¹/₈ inch wide. Lawns and disturbed ground. Spring–summer.

V. peregrina, known as Purslane Speedwell, is smooth and has narrowly oblong leaves about 1 inch long and tiny white flowers. Moist fields and gardens. Spring–summer.

Gratiola spp., Hedge Hyssop. The flowers of the Hedge Hyssops appear to have a 4-lobed irregular corolla, but the upper lobe actually consists of two that are fused except at their very summits. The calyx is 5-lobed and subtended by 2 sepal-like bracts. There are 2 fertile stamens, no sterile ones. These are plants of wet woods, marshes, and ditches, flowering between late spring and fall.

Gratiola neglecta (plate 456) is less than 1 foot high, with finely toothed elliptic leaves. Its flowers are ³/₈ inch long, single in the leaf axils on slender pedicels about twice as long, with a yellow tube and white lobes. (The others are white or purplish.)

G. viscidula grows to 2 feet tall and is partially sticky-glandular, with toothed ovate leaves. The flowers are on ⁵/₈-inch pedicels.

G. virginiana is nearly as tall and has elliptical leaves and flowers on stout but very short pedicels.

G. pilosa is hairy, with its flowers sessile in the axils of ovate leaves. The corolla is small, only about $1/4$ inch long.

Lindernia dubia (incl. **L. anagallidea**), False Pimpernel (plate 457). Our species have opposite leaves and flowers less than $3/8$ inch, the upper lip small and the lower with three spreading lobes. There are no bracts below the calyx as in *Gratiola*. There are 4 stamens, 2 of which lack anthers. *L. dubia* has ovate leaves up to 1 inch long and pale purple flowers on long stalks in the axils. Wet or muddy soil. Spring–fall.

L. monticola (plate 458) has a basal rosette of $1/2$-inch elliptic leaves and several slender, erect stems bearing minute bracts. The flowers are violet, pale, and streaked within. Wet soil near rock outcrops. Spring–summer.

L. saxicola is a rare species found only on rocks in mountain streams. It is only 4 inches high, and has cauline leaves $1/4$ to $3/8$ inch long. Summer.

Buchnera americana, Bluehearts (plate 459). This is an erect, unbranched, rough-hairy plant with coarsely toothed opposite, sessile leaves. It terminates in an open, elongate spike of blue-purple salverform flowers with a slender tube $5/8$ inch long and 5 shorter, almost regular, spreading lobes. Open areas. Summer-fall.

Bacopa caroliniana, Water Hyssop (plate 460), is another plant with almost regular bilabiate flowers. It is a creeping or even floating plant, and its flowers are borne on short stalks singly in the axils of the upper leaves, which are ovate and entire. They are blue, about $5/8$ inch long, bell-shaped with 5 lobes (the 2 upper ones are joined for half their length). Wet shores and shallow water. Summer–fall.

Mazus japonica (plate 461). Introduced from Asia, *Mazus* is frequently found growing in lawns. It is a creeping plant with small, obovate basal leaves and $1/4$-inch, lavender-blue bilabiate flowers, the upper lip much smaller than the lower. Spring–summer.

BIGNONIACEAE, BIGNONIA FAMILY

Bignonia capreolata (*Anisostichus capreolata*), Cross Vine (plate 462). The large colorful blossoms of this woody vine and the next would seem reason enough to include them in a wildflower book. Cross Vine has its leaves in opposite pairs; each consists of two entire leaflets and a branched tendril equipped with adhesive disks which enable the plant to climb to considerable heights. In the axils there are clusters of widely tubular flowers 2 inches long with 5 short, irregular spreading lobes, dull red outside, yellow within. The common name alludes to its cross-shaped pith. Woods, thickets, and streamsides. Spring.

Campsis radicans, Trumpet Creeper (plate 463), lacks tendrils but climbs by means of aerial roots. Its leaves are pinnately compound with as many as 15 ovate, sharply toothed leaflets 1 $1/2$ to 3 inches long. The orange-red flowers are narrowly campanulate with 5 flaring, slightly irregular lobes. Margins of woods, fencerows, and roadsides, sometimes becoming weedy. Summer.

Trumpet Creeper is widely cultivated as a showy ornamental, but it is well to note that in some locales it is also known by the name "Cow Itch Vine" to susceptible individuals who develop a rash upon contact with it.

OROBANCHACEAE, BROOM RAPE FAMILY

Conopholis americana, Squaw Root (plate 464). The Broom Rape Family is made up of flowering plants that lack green color and parasitize various other plants. The stem of Squaw Root is an inch or so thick, up to 8 inches high, and unbranched, but it often grows in crowded clusters. It is yellow, rapidly turning brownish with age, and is covered by overlapping scaly bracts. A $1/2$-inch whitish 2-lipped flower projects from the stem above each bract. Spring.

True to its name, Beech Drops, **Epifagus virginiana** (plate 465), parasitizes the roots of beech trees. It is slender and usually branched. The upper flowers are $3/8$ inch long, tubular, 2-lipped with 4 short lobes; color is variable, but in most plants they are either yellowish or purplish. The lower flowers are much smaller and cleistogamous. Summer–fall.

Orobanche uniflora, One-flowered Cancer Root (plate 466), sends up one to many fleshy stalks from an underground stem, each terminating in a solitary, fragrant, white or pale violet flower. The corolla is tubular and slightly curved, ³/₄ inch long, with 5 spreading lobes. Rich woods. Spring.

ACANTHACEAE, ACANTHUS FAMILY

Justicia americana, Water Willow (plate 467). This plant gets its common name from the slender, lanceolate, willow-like leaves. The flowers, about ³/₄ inch across, are white to lavender with purple markings. They are 2-lipped, the upper notched and the lower divided into 3 lobes with the lateral ones widely divergent. Shallow water, wet shores, and sandbars. Summer–fall.

***Ruellia* spp.,** Wild Petunia. The *Ruellias* have lavender-blue trumpet-shaped flowers up to 2 inches long, with 5 flaring lobes, growing in the axils of the paired leaves. They grow in dry woods, blooming in spring and summer. Variability and natural hybridization make the identification of individual species difficult.

Ruellia caroliniensis (*R. ciliosa*) (plate 468), our most common representative of the genus, is a somewhat hairy plant with stalked leaves. It bears its flowers in sessile clusters (very few are open at one time) in the upper axils, and the calyx teeth are linear.

R. humilis is similar but has sessile leaves. ***R. purshiana*** bears its flowers singly on stalks arising from the median axils.

R. strepens is essentially smooth, has short-petioled leaves, and differs markedly from the others in having lanceolate calyx lobes.

PHRYMACEAE, LOPSEED FAMILY

Phryma leptostachya, Lopseed (plate 469). *Phryma* grows from 1 to 3 feet high and has opposite, petioled, coarsely toothed ovate leaves up to 6 inches long. Its flowers are bilaterally symmetric, purplish, ¹/₄ inch long, in regularly spaced pairs on one or more slender spikes. They are horizontal but when wilted hang downward, hence the name Lopseed. Woods. Summer.

RUBIACEAE, MADDER FAMILY

***Galium* spp.,** Bedstraw. Bedstraws can be recognized by the whorls of usually narrow leaves at intervals along their slender, square stems, and by their small rotate flowers with 4 (rarely 3) lobes. Plants with leaves in 6's or 8's will sometimes have whorls containing 1 or 2 fewer leaves. They flower at various times from spring until late summer.

> Leaves in whorls of 8: G. *aparine*, G. *mollugo*, G. *verum*
> Leaves in whorls of 6: G. *asprellum*, G. *concinnum*,
>   G. *triflorum*, G. *tinctorium*
> Leaves in whorls of 4: G. *boreale*, G. *pilosum*, G. *obtusum*,
>   G. *circaezans*, G. *lanceolatum*, G. *pedemontanum*,
>   G. *latifolium*, G. *hispidulum*

***Galium aparine*,** Cleavers, reclines on other vegetation, supporting itself with down-curving prickles. Its principal leaves are narrowly oblanceolate, up to 3 inches long. Long peduncles arise from the leaf axils, each bearing a few white flowers. Thickets.

***G. mollugo*,** Wild Madder, is erect and without prickles. Its leaves are also narrowly oblanceolate but not more than ³/₄ inch long. The white flowers are numerous in a loose, forked panicle. Fields and roadsides.

Yellow Bedstraw, ***G. verum*** (plate 470), is another smooth, erect plant. It has linear leaves up to 1 inch long and dense, stalked clusters of yellow flowers from the upper axils, forming a large, many-flowered panicle. Fields and roadsides.

***G. asprellum*,** well-named Rough Bedstraw (plate 471), is a large branched, reclining plant with recurved prickles on the stem and leaves. The principal leaves are narrowly elliptic, up to ³/₄ inch long, and the flowers are white. The inflorescence is a loose panicle of many few-flowered clusters Thickets.

Shining Bedstraw, **G. concinnum,** is also reclining and branched, with an open, white-flowered inflorescence, but it is essentially smooth. Its leaves are linear. Dry woods.

Sweet-scented Bedstraw, **G. triflorum,** is another smooth, prostrate species. Its flowers are greenish white and are in groups of 3 on long axillary and terminal peduncles.

G. tinctorium is our only Bedstraw with 3-lobed flowers. They are white, very small, and in clusters of 3. This is a weak, much-branched species with narrow leaves. Wet soil and swamps.

Northern Bedstraw, **G. boreale,** is erect, smooth, with linear-lanceolate leaves mostly less than $^3/_{16}$ inch wide. The inflorescence is a large panicle with very numerous, bright white flowers. Open areas.

Hairy Bedstraw, **G. pilosum,** is also erect but is invested with straight hairs and has oval leaves. The flowers are stalked at the ends of many branchlets and may be greenish white or purplish. Dry woods.

G. obtusum is matted, smooth, and diffusely branched, with white flowers and lanceolate leaves up to 1 inch long, rounded at the apex. Moist areas.

In Wild Licorice, **G. circaezans,** the leaves are oval or elliptic, broadest at the middle and blunt at the apex, up to 1 $^1/_4$ inches long and half as wide. The greenish flowers are sessile and widely spaced on forked branches from the upper axils. Dry woods. **G. lanceolatum** is similar but has lanceolate leaves up to 2 inches long, broadest near the base. Its flowers are yellowish, turning purple.

G. pedemontanum is a hairy plant with elliptic leaves less than $^3/_8$ inch long and very small yellow flowers. Fields and waste areas.

G. latifolium (plate 472) is erect and branched, with ovate-lanceolate leaves up to 2 $^1/_4$ inches long and $^1/_2$ to $^3/_4$ inch wide, tapering to a sharp point at the end. The flowers are purple or brownish maroon. Rich woods.

G. hispidulum is unique among our species in having purple, succulent berries instead of dry fruits. It is diffusely branched, with rough hairs and white flowers. Dry soil.

Sherardia arvensis, Field Madder. Except for the flower color and arrangement, this diffuse, spreading, rough-hairy plant with whorled leaves might easily be taken for a Bedstraw. The flowers are blue or pink with a slender tube and 4 spreading lobes. They are in terminal heads surrounded by an involucre of 8 leaves joined at their bases. The stem leaves are narrow, ⁵/₈ inch long, and mostly in whorls of 6. Waste places. Spring–summer.

Houstonia caerulea (*Hedyotis caerulea*), Bluets, Quaker Ladies, Innocence. Our species of *Houstonia* have opposite leaves and small 4-merous stalked flowers. The first three are delicate plants with solitary salverform flowers. *H. caerulea* is erect, rising from a basal rosette of spatulate leaves ¹/₂ inch long. The corolla is about ¹/₂ inch wide, blue to white with a yellow eye. Stream banks, clearings, and grassy areas. Spring–summer.

Thyme-leaved Bluets, **H. serpyllifolia** (*Hedyotis michauxii*) (plate 473), have prostrate stems with numerous roundish leaves under ¹/₄ inch. The flowers are ¹/₂ inch wide, bright blue or occasionally white with a prominent yellow eye. Moist woods, seeps, and stream banks. Spring–summer.

Small Bluets, **H. pusilla** (*Hedyotis crassifolia*) (plate 474), are similar but only 2 to 4 inches high, and the flowers are ³/₈ inch across, lilac with a reddish eye. Open woods and meadows. Spring.

H. purpurea (*Hedyotis purpurea*), Purple or Summer Bluets (plate 475). This and the remaining species have funnelform flowers in clusters. In *H. purpurea* the leaves are ovate-lanceolate and sessile. The numerous flowers, which are up to ³/₈ inch long, vary from light purple to white. Woodlands and roadsides. Spring–summer. Mountain Bluets, **H. montana,** are like *H. purpurea* but have deep purple flowers and shorter, ovate leaves. Restricted to rocky outcrops at high elevations.

H. longifolia (*Hedyotis longifolia*) and **H. tenuifolia** (*Hedyotis tenuifolia*) also resemble *H. purpurea*. The first has leaves slightly more than ¹/₈ inch wide and short-pediceled flowers. The second is very slender, with leaves under ¹/₈ inch wide and longer pedicels. Both inhabit dry soils. Summer.

H. canadensis (*Hedyotis canadensis*) has narrow leaves with ciliate margins, and is unique among these five in having a rosette of numerous leaves present at time of flowering. Spring.

In Houstonia it is possible to see dimorphic flowers, in which some have short stamens and a long pistil and others just the reverse. This arrangement caught the attention of Charles Darwin, who concluded that it has the effect of making self-fertilization virtually impossible.

Diodia teres, Buttonweed (plate 476), is a low herb with stiff, linear, sessile leaves mostly 1 inch long, and pink 4-lobed funnelform flowers ¹/₄ inch long, borne 1 or 2 in the axils. There are 4 very short sepals, and stipules with erect bristles as long as the flowers. The stigma is undivided. Dry fields and waste places. Summer–fall.

D. virginiana (plate 477) has thin, narrowly lanceolate leaves up to 2 inches long, and solitary white 4-lobed salverform flowers with a slender tube about ³/₈ inch long. There are 2 linear sepals, and the style has 2 filiform stigmas. Wet locations. Summer–fall.

Mitchella repens, Partridge Berry (plate 478), is an evergreen trailing herb with opposite, ovate leaves about ¹/₂ inch long, and pairs of white or pinkish flowers at the ends of branches. The corollas are funnelform with 4 spreading lobes, about ¹/₂ inch long, and densely hairy within. The ovaries are fused, resulting in a red twin berry showing the remains of 2 calyxes. Rich woods. Late spring–summer.

CAPRIFOLIACEAE, HONEYSUCKLE FAMILY

Triosteum spp. Tinkerweed is just one of the names given to plants of the genus *Triosteum*; others are Horse Gentian, Feverwort, and Wild Coffee. All our species are coarse plants with pairs of large entire leaves and $^1/_2$ to $^3/_4$-inch funnel-shaped flowers with 5 erect lobes, growing sessile in the axils. They grow in open woods and clearings, and flower in spring. Intermediate forms sometimes make it difficult to separate the species.

Triosteum aurantiacum (plate 479) has ovate leaves more than 2 inches wide, sessile but not connate. The flowers are purplish red and the fruits are hairy yellow drupes.

T. angustifolium has narrower, sessile, lanceolate leaves. Its flowers are pale yellow and the drupes orange. Both species have rough hairs on the stems.

T. perfoliatum has connate-perfoliate leaves at least in the middle, and the herbage is soft gray-hairy. Its flowers may be purplish red or greenish yellow; the drupes are red-orange.

Valerianaceae, Valerian Family

Valerianella locusta (*V. olitoria*), Corn Salad. Our species of *Valerianella* are succulent, dichotomously forked plants with opposite, simple leaves on the stem and tiny 5-lobed flowers with 3 stamens, crowded into terminal heads. They are usually found in fields and other disturbed areas, blooming in spring. *V. locusta* is under 1 foot high. Its flowers are pale blue, only $^1/_{16}$ inch long, the stamens not appreciably protruding.

Two of our species have white flowers with conspicuously exserted stamens. In **V. radiata** (plate 480) they are about the same size as those of *V. locusta* and are borne in densely flowered cymes, but in **V. umbilicata** they have somewhat larger corollas and are in open, loosely flowered clusters.

Dipsacaceae, Teasel Family

Dipsacus sylvestris (*D. fullonum*), Teasel (plate 481). Teasel is more noticeable during the winter, when its spent flower heads have dried to a dark brown and punctuate old fields and roadsides. It is a stout, erect plant, usually 5 feet high, with opposite connate leaves, lanceolate, and only slightly toothed if at all. The inflorescence is an egg-shaped spike 2 inches or more in length, packed with tiny lavender flowers that often begin opening about midpoint, then progress upward and downward. Very conspicuous except where hidden by the flowers are many bristlelike awns. There also are a number of curved linear involucral bracts, some of which may extend above the head. Summer–fall.

When dry, the spiky flower heads of Teasel proved uniquely useful for raising the nap on woven fabric since, unlike metal, the spines would break before damaging the cloth. The specific epithet fullonum *was a misnomer, implying erroneously that the heads were used for "fulling" cloth, which is a different process by which it is made thicker.*

CUCURBITACEAE, GOURD FAMILY

Sicyos angulatus, Bur Cucumber (plate 482). Our representatives of this family are climbing vines with tendrils and axillary inflorescences. The principal leaves are palmately lobed and finely toothed. *Sicyos angulatus* has branched tendrils, and white flowers with 5-lobed rotate corollas ³/₈ inch wide, in stalked clusters. Thickets and disturbed areas. Summer–fall.

Echinocystis lobata, known as Wild Cucumber or Balsam Apple (plate 483), also has branched tendrils. Its flowers are white, ¹/₂ inch wide, and have 6 narrow lobes. The staminate flowers are disposed in long panicles each with a solitary pistillate flower near its base. Thickets. Summer–fall.

Melothria pendula, Creeping Cucumber (plate 484), shares a number of characters with the two preceding species. However, the flowers are yellow, bell-shaped, 5-lobed, and very small—the pistillate ¹/₄ inch wide and the staminate even smaller. Also, the tendrils are unbranched. Woods and thickets. Summer–fall.

CAMPANULACEAE, HAREBELL FAMILY

Campanula americana, Tall Bellflower (plate 485). This is our only species of *Campanula* that does not have a bell-shaped flower as its names would seem to call for. The corolla is rotate, measuring about 1 inch across the 5 flat, light blue segments; the long style curves downward, then up at the tip. It is an erect plant up to 6 feet high, with lanceolate toothed leaves, terminating in a long spike. Thickets and moist woodland margins. Summer–fall.

European Bellflower, **C.** *rapunculoides,* is a garden escape, with unbranched stems up to 3 feet from a creeping rhizome. The lower leaves are ovate, stalked, and irregularly toothed. Its flowers are blue, narrowly campanulate, 1 inch or longer, nodding on very short pedicels in a one-sided raceme. Disturbed or waste ground. Summer.

Southern Harebell, **C.** *divaricata* (plate 486), is extremely delicate and diffusely branched and has numerous small flowers nodding on slender petioles in loose panicles. They are about ¼ inch long, blue, campanulate with recurved lobes and a straight style that protrudes conspicuously beyond the corolla. The leaves are lanceolate with relatively large, sharp teeth. Rocky slopes. Summer–fall.

C. *aparinoides,* Marsh or Bedstraw Bellflower, inhabits damp meadows and other wet areas. It is a weak, reclining plant with slightly scabrous stems and narrowly lanceolate leaves. The funnelform flowers are ¼ to ⅜ inch long, white or pale blue, and are solitary on ascending pedicels. Summer.

Lobelia **spp.** Our Lobelias have bilaterally symmetric flowers in which the lower lip is divided into 3 spreading lobes and the upper into 2 smaller, erect lobes. Through a cleft in the top of the corolla tube there projects a column in which the stamens are joined around the style.

Lobelia siphilitica, Great Blue Lobelia (plate 487), is a mostly smooth, stout plant with lanceolate leaves and a raceme of blue flowers 1 inch long. The lower lobes are deflexed, and the corolla tube is somewhat inflated and has several darker blue lines, especially on the underside. Wet ground. Late summer–fall.

Downy Lobelia, **L.** *puberula* (plate 488), has a fine-hairy stem, toothed leaves, and a one-sided raceme of purplish blue flowers about $^3/_4$ inch long. They lack the inflated tube and striping of *L. siphilitica*. Open woods. Summer–fall. Two very similar species are recognized on the basis of characters some of which are less consistent than one might desire. **L. amoena** is essentially smooth and may have untoothed leaves. **L. glandulosa,** which is much less common in the mountains, has narrow leaves, a few-flowered raceme, and may have callous-tipped teeth on the calyx.

L. inflata, Indian Tobacco (plate 489). The flowers of Indian Tobacco are pale violet or blue and only $^3/_8$ inch long; the calyx becomes much inflated. They are borne singly in the axils of a leafy raceme. The stem is hairy and often branched, and the median leaves are ovate. Fields, roadsides, and open woods. Summer–fall.

Early explorers not only were introduced to the native tobacco (Nicotiana rustica) *by the Cherokees but observed them smoking the dried leaves of* Lobelia inflata *as well, and referred to it as Indian Tobacco. The plant has been found to contain the alkaloid lobeline, which is similar to the nicotine in tobacco and is used in anti-smoking preparations.*

L. spicata, Pale Spike Lobelia. This is an unbranched, usually smooth plant with $^3/_8$-inch pale blue to white flowers in a crowded raceme. Its leaves are obovate and mostly near the base; the upper ones are reduced to bracts. Meadows and thickets. Summer.

Nuttall's Lobelia, **L. nuttallii,** is very slender and has narrow, linear leaves. The flowers are $^3/_8$ inch long and light blue with a white center and two greenish spots on the lower lip. They are stalked and widely spaced in a raceme. Moist areas. Summer–fall.

Cardinal Flower, **L. cardinalis** (plate 490). This is our largest Lobelia and the only one with red flowers, and certainly it is one of the most striking of all wildflowers. A distinctive feature of the vibrant scarlet, 1 $^1/_2$-inch flowers is the long, curved filament tube which ascends far beyond the upper corolla lobes. Wet places and stream banks. Late summer–fall.

Triodanis perfoliata (*Specularia perfoliata*), Venus' Looking Glass (plate 491). This is an unbranched plant with an angled stem and clasping or sessile leaves less than 1 inch long and about as wide. Solitary violet, 5-lobed rotate flowers ¹/₂ inch across are tucked into the upper leaf axils (cleistogamous flowers are borne in the lower axils). Dry fields and clearings. Spring–summer.

The name "Venus' Looking Glass" (as well as the former generic name Specularia, *which generated it and was in turn derived from the Latin term for "mirror") seems rather far-fetched since all it refers to is the fact that the tiny flattish seeds appear shiny as though polished.*

ASTERACEAE, SUNFLOWER FAMILY

With the possible exception of the Orchidaceae this is the largest family of plants in the world, and certainly includes many more of our southern mountain wildflowers than does any other. As a group, they are commonly called the Composites—a reference to their unique characteristic of having multiple flowers aggregated into dense clusters that simulate single flowers. Because this mimicry can lead to erroneous perceptions, a small but important lexicon has been formulated for describing these plants, and a clear understanding of what is meant by these terms (which are italicized here) is a prerequisite to accurate identification.

What may appear to be a flower is actually a collection of small individual flowers (sometimes called *florets*) inserted on the expanded summit of the stalk (the *receptacle*) and known as a *head*. Each floret is either of two types. Those of the first kind have tubular corollas with short lobes or teeth (usually 5) at the apex, and are known as *disk flowers*. The others are flat (except at the very base) and may or may not have vestigial teeth at the apex; these are called *ray flowers*. A head may consist solely of tubular flowers (a *discoid head*), or may have only ray flowers (a *ligulate head*), or may contain both tubular flowers crowded into a central disk and ray flowers extending outward from the perimeter, in which case it is a *radiate head*. (Familiar examples of discoid, ligulate, and radiate heads are Thistles, Dandelions, and Daisies, respectively.)

A floret may have stamens (5) or a pistil (with a 2-cleft style if it is functional), or both or neither. A feature that is sometimes diagnostically significant is the *pappus* (which takes the place of a calyx); this is represented by bristles, awns, hairs, scales, etc., arising from the summit of the fruit (the *achene*), or from the end of a slender extension of the achene called the *beak* if such is present (see *Taraxacum officinale*, plate 502).

The lower portion of the head is surrounded by a cup-like *involucre* made up of small, usually green, leaflike *bracts* which are arranged in one or more series and often overlap. In some cases the individual florets are themselves subtended by a different kind of bract—thin or papery and sometimes bristly; to avoid confusion with those of the involucre, these receptacular bracts are referred to as the *chaff*.

In the following descriptions, the width of a flower head should be understood as meaning its diameter measured between the apexes of extreme opposite florets.

> Heads ligulate: *Prenanthes, Cichorium, Lactuca, Tragopogon, Hieracium, Taraxacum, Krigia, Hypochoeris, Pyrrhopappus, Sonchus, Crepis, Lapsana, Picris*

> Heads discoid; flowers white, yellow or green: *Antennaria, Anaphalis, Gnaphalium, Mikania, Ageratina, Eupatorium* (part), *Kuhnia, Erechtites, Arnoglossum, Synosma, Rugelia, Senecio* (part), *Tanacetum, Matricaria, Bidens* (part)

> Heads discoid; flowers pink, purple or blue: *Cirsium, Carduus, Arctium, Centaurea, Elephantopus, Pluchea, Gamochaeta, Liatris, Vernonia, Eupatorium* (part), *Marshallia*

> Heads radiate; rays yellow: *Rudbeckia, Ratibida, Helenium, Helianthus, Heliopsis, Tetragonotheca, Coreopsis, Bidens* (part), *Pityopsis, Chrysopsis, Heterotheca, Senecio* (part), *Chrysogonum, Silphium, Verbesina* (part), *Smallanthus, Inula, Tussilago, Solidago, Euthamia*

> Heads radiate; rays not yellow: *Echinacea, Polymnia, Leucanthemum, Anthemis, Bellis, Aster, Erigeron, Verbesina* (part), *Galinsoga, Parthenium, Achillea, Solidago* (part), *Conyza, Eclipta*

Heads ligulate

Prenanthes **spp.** Our plants in this genus are erect, with petioled leaves, toothed or lobed, generally triangular in outline but otherwise exceedingly variable in shape, blooming in woods and on road banks during the late summer and fall. The heads are $^1/_2$ inch long, bell-shaped, and pendent, in panicles; the involucre consists of a series of long principal (inner) bracts with several much smaller (outer) bracts at the base. As with all of the other plants in this section, the stems exude a milky juice.

Prenanthes altissima, Tall White Lettuce. In this species there are usually 5 florets in each head, subtended by 5 smooth principal bracts. The rays are greenish white, and the pappus is made up of long creamy to yellowish tan hairs.

The less common **P. roanensis,** Rattlesnake Root (plate 492), of high elevations also has few-flowered heads, but the rays are yellowish, the involucres have some long spreading hairs and black-tipped bracts, and the pappus is whitish.

P. alba, White Lettuce, Rattlesnake Root. In our other three species, each head contains 8 or more flowers and the involucres usually have 6 to 8 principal bracts. *P. alba* has a glaucous stem and flowers with whitish rays, a smooth involucre, and cinnamon-brown pappus.

P. trifoliata, Gall-of-the-Earth, is smooth-stemmed but not glaucous. The rays are yellowish white and the pappus straw-colored. The outer bracts are short and triangular.

P. serpentaria, Lion's-foot, is somewhat similar to the preceding, but the inner bracts are minutely bristly and the outer bracts are narrowly lanceolate. Also, the leaves tend to be thicker and the lobes blunt rather than pointed as in *P. trifoliata.*

Cichorium intybus, Chicory, Blue Sailors (plate 493). The soft blue color of this attractive European immigrant is unique among our composites. The principal leaves are pinnately cleft and form a basal rosette. From this arises a stiff stem bearing 1 to 1 $^1/_2$-inch heads in the axils of much smaller clasping leaves. The petal-like rays (which occasionally vary to white) are distinctly 5-lobed at the truncate apex. Roadsides, fields, and waste places. Late spring until frost.

In the Old World, an infusion made from the roasted and ground taproot of Chicory was used as a substitute for coffee. The bitter taste of this beverage is echoed in the leaves of a close relative, Cichorium endivia, which we know as Endive.

Lactuca serriola var. **integrata** (*L. scariola*), Prickly Lettuce (plate 494). Our wild species of *Lactuca* are tall plants with variable cauline leaves and open panicles of small flower heads, occurring in woodland borders and various open habitats and flowering in summer and fall. Prickly Lettuce is mostly smooth-stemmed and derives its common name from the copious fine prickles on the margins and midrib of its leaves, which in this variety are oblong or oblanceolate and sagittate-clasping. The heads are small and yellow-flowered.

L. saligna, Willow-leaved Lettuce, is also smooth and has conspicuously sagittate-based linear to lanceolate leaves, which are either entire or slightly toothed, and sparsely if at all prickly. Its flowers are yellow.

L. hirsuta is a somewhat hairy plant with leaves that are pinnately dissected and do not clasp the stem. The yellow flowers are in relatively large heads, the involucres measuring $^1/_2$ inch or more in height.

L. canadensis is smooth and sometimes glaucous. Its leaves are extremely variable, ranging from pinnatifid to entire, but without spiny margins. There are many pale to deep yellow flowers in each head, but the involucres are less than $^1/_2$ inch high.

L. biennis, Tall Blue Lettuce. Typically this species has blue flowers, but in one form they are yellow. The individual florets are very numerous; the pappus is light brown. The panicle is elongate, its branches short and ascending. The stem is green and the leaves are auriculate-clasping at the base.

L. floridana, known as Woodland Lettuce (plate 495), has blue flowers but fewer (11 to 17 to a head), and the pappus is white. The stems are sometimes reddish and the leaves not clasping. The panicle is diffuse, with long, spreading branches.

L. graminifolia has mostly basal leaves, up to 1 foot long but less than ¹/₄ inch wide, frequently lobed. The blue to violet flowers are borne in a widely spreading, diffuse panicle. This is primarily a coastal species and blooms in spring and summer.

Tragopogon spp. Our species of *Tragopogon* have been introduced from the Old World and are found in fields and waste areas, flowering in the morning hours during spring and summer. They have sheathing, grasslike foliage and large terminal flower heads. The tawny pappus is plumose and creates globose seed heads resembling those of the Dandelions but much larger.

Tragopogon porrifolius, Salsify, Oyster Plant (plate 496), is cultivated for its edible root. The rays are purple and are exceeded by the narrow involucral bracts, which are 1 to 1 ¹/₂ inches long.

T. dubius (*T. major*) is similar but with yellow rays. The flower stalk is gradually but conspicuously broadened beneath the involucre. **T. pratensis,** Yellow Goat's Beard, resembles it but its stalk is not swollen, and the involucral bracts are shorter than the rays.

Hieracium spp., Hawkweed. Along with the other remaining genera in this section, the *Hieraciums* have yellow or orange dandelion-like flowers. They are found mostly in fields, along roadsides, or in open woods, and bloom from late spring or summer until early fall.

Hieracium pilosella, Mouse-ear Hawkweed (plate 497), is a small stoloniferous plant forming extensive mats. It has a basal rosette of oblanceolate leaves bearing long bristles and a hairy scape with a solitary 1 inch head of lemon-yellow ray flowers.

H. aurantiacum, Devil's Paintbrush (plate 498), is rather similar, but has slightly smaller, brilliant red-orange flower heads, several in a compact cluster. Blackish glandular hairs are often seen in both species.

In England the black specks on this colorful little weed brought to mind the spattering of coal dust which is the miner's lot, and gave it the name "Grimm the Collier."

H. caespitosum (*H. pratense*), King Devil, Field Hawkweed (plate 499). This and the remaining species of *Hieracium* are taller plants with yellow flowers in heads that are usually ¹/₂ to ³/₄ inch wide. *H. caespitosum* has narrowly elliptic leaves, long-hairy on both sides, all basal except for 1 or 2 much smaller ones on the stem. There are several heads in a compact inflorescence. The stem and involucres have conspicuous black-tipped glandular hairs. **H. florentinum,** Smooth Hawkweed, is similar but essentially smooth, except for glandular hairs on the involucres and sparse hairs on the leaf margins.

H. venosum, Rattlesnake Weed (plate 500), is a slender, smooth-stemmed plant with a basal rosette of leaves which have prominent, usually purple veins. The flower heads are in a loose, open panicle.

H. scabrum, Rough Hawkweed, has a rough-bristly stem. Its basal leaves are usually absent at flowering time; the cauline ones are progressively reduced upward and extend to the flat-topped inflorescence. Each head bears more than 40 flowers.

In **H. gronovii,** Hairy Hawkweed, the basal leaves are usually persistent, and the cauline leaves are restricted to the lower half of the stem. The inflorescence is elongate but open, and the heads have fewer than 40 flowers each.

H. paniculatum, Panicled Hawkweed, is a slender plant, smooth in the inflorescence, which is an open panicle of long, spreading, flexuous branches. Its leaves are cauline, entire or with distant teeth. The flower heads are small (¹/₂ inch wide or less).

Taraxacum officinale, Common Dandelion (plate 501 and plate 502), is doubtless our best-known weed. Its leaves, which are cut into sharp, backward-pointing lobes, are all at the base, and each hollow scape bears a single flower head. The bracts of the involucre are in 2 rows and reflexed. The familiar seed heads are formed by the white bristles of the pappus, which are attached at the summit of the long beaks of the greenish brown achenes.

The smaller **T. laevigatum** (*T. erythrospermum*), Red-seeded Dandelion, is more scattered in its distribution. It can be distinguished from the common species by its more deeply cut leaves, short spreading outer bracts, and at maturity by the reddish achenes.

Krigia spp., Dwarf Dandelion. *Krigias* differ from the true Dandelions in that the involucral bracts are in a single series and the achenes are beakless. They bloom in spring and summer.

Krigia virginica is leafy only at or near the base, the leaves usually pinnatifid or at least toothed. There are several orange or yellow flower heads with rays $^1/_4$ inch long. The pappus has less than 10 bristles. Open spaces.

K. oppositifolia (*Serinia oppositifolia*) is another small-flowered plant but has both basal and cauline leaves, the latter sometimes appearing opposite. The stem is branched, and there is no pappus. Moist, low places.

K. biflora, known as Cynthia, is taller. Its basal leaves are elliptic-oblanceolate, $^3/_4$ inch or more wide, and either entire or with teeth or lobes near the tapered base; there are also a few small clasping leaves on the stem. The heads are orange with $^1/_2$-inch rays, and are borne on long stalks, usually several from an axil. In this and the remaining species, the pappus has numerous bristles. Woodlands and fields.

In **K. montana,** Mountain Cynthia (plate 503), the leaves are narrower and may be entire or coarsely toothed near the base. The flowers are similar to those of *K. biflora* but are bright yellow and solitary on leafless stalks. Wet cliffs and rocky stream banks.

K. dandelion has glaucous, narrow, entire to dentate leaves, all in a basal rosette. Its flowers are solitary on a long scape and relatively large, the rays about $^3/_4$ inch long. Open places.

Hypochoeris radicata, Cat's Ear (plate 504). The leaves of this widespread weed are in a basal rosette, hairy on both sides, and pinnately cleft with rounded lobes. The smooth stem is sparingly branched, with only a few terminal flower heads, each about 1 inch wide and bright yellow. The bracts are in several overlapping rows, and the pappus is pale tan and plumose. Fields and waste places. Spring–summer.

Pyrrhopappus carolinianus, False Dandelion (plate 505), is smooth and has long, narrow, toothed to pinnatifid leaves, those on the stem well developed at time of flowering. There are a few 1 ¹/₂-inch, light yellow flower heads on long stalks. The pappus is dull reddish tan and is subtended by a ring of white hairs. Fields and waste areas. Spring–summer.

Sonchus arvensis, Field Sow Thistle. The Sow Thistles are large, coarse weeds with prickly-margined leaves and heads of yellow flowers blooming in summer and fall. In *S. arvensis* the lower leaves are pinnately lobed, the upper ones less so and with small auriculate-clasping bases. The flower heads are 1 to 1 ¹/₂ inches wide.

S. oleraceus, Common Sow Thistle, also has deeply lobed leaves, the lowermost ones with a large terminal segment. Most are conspicuously sagittate-clasping with pointed basal lobes. Its flower heads are smaller, averaging ³/₄ inch across.

S. asper, Spiny-leaved Sow Thistle (plate 506), is similar except that its leaves are frequently unlobed, and the clasping bases are broadly rounded.

Crepis pulchra, Small-flowered Hawk's Beard, has oblong, coarsely toothed leaves, mostly confined to the lower half of the plant. The panicle is openly branched, and the yellow rays are about ¹/₄ inch long. The inner bracts of the involucre are hairless, have a thickened midrib, and are subtended by a ring of very short outer bracts. Fields and waste areas. Spring–fall.

In **C. capillaris,** Smooth Hawk's Beard (plate 507), the basal leaves are often pinnatifid and the cauline ones narrow with pointed sagittate-clasping basal lobes. The inner bracts are hairy and not keeled, and the outer bracts are half as long. The rays are ³/₈ inch long. Spring–summer.

Lapsana communis, Nipplewort (plate 508), is a slender plant, branched above, with an open inflorescence of $^1/_4$- to $^1/_2$-inch heads each with a dozen or fewer yellow flowers. The leaves are stalked, long-ovate, and toothed, the lower ones larger and often bearing a few small lobes on the petiole. The involucre has one series of long, narrow inner bracts and a ring of minute outer ones; there is no pappus. Woods, fields, and waste areas. Summer.

Picris hieracioides, Bitterweed, is a coarse plant with hairs that are forked at the apex. Its leaves are shallowly toothed or lobed, the basal ones often absent and the cauline ones sessile or clasping. The involucral bracts are in several overlapping series, and the pappus bristles are white and plumose. Its flowers are yellow. Fields and waste areas. Spring–fall.

Heads discoid; flowers white, yellow or green
(See also Bidens and Senecio)

Antennaria **spp.**, Pussytoes. The *Antennarias* are stoloniferous plants with mostly basal leaves, often forming mats, and appear woolly or silky due to a covering of white hairs. Usually the heads of pistillate flowers (recognizable by their bifid styles) are borne on different plants than the staminate heads (which have conspicuous white-tipped bracts and often undivided styles). They bloom in spring, in woods or fields.

Antennaria solitaria, Solitary Pussytoes (plate 509), is unique in having only one head per plant, terminating a short stem with small, bractlike leaves. The obovate basal leaves usually have 3 or 5 main nerves.

The other species, which bear several heads in a terminal cluster, exhibit many variable characteristics and are difficult to identify. **A.** *plantaginifolia,* Plantain-leaved Pussytoes (plate 510), is a taller plant with long-stalked, obovate basal leaves with 3 or more principal nerves; those on the stem are narrow, much reduced in size, and ascending.

A. *neglecta* is one of a complex group known as Field Pussytoes. These are shorter plants with smaller, narrower basal leaves which usually are 1-nerved and with-

out distinct petioles. One exception to this last characteristic is **A. virginica,** sometimes called Shale Barren
Pussytoes.

Anaphalis margaritacea (plate 511) is called Pearly
Everlasting because it retains its general appearance
indefinitely in a dried state. It is a dioecious, white-
woolly plant of a generally northern distribution, in-
habiting dry, open places. It bears a resemblance to the
Antennarias, but has spreading linear cauline leaves with
revolute margins and blooms in late summer. The flower
heads are round with pearly white petaloid bracts and
are crowded in a short inflorescence.

Gnaphalium obtusifolium, Sweet Everlasting, Rabbit
Tobacco, Catfoot. In this genus the central florets in
each head have both stamens and pistils. G. *obtusifolium*
has lanceolate cauline leaves, a white-woolly stem, and
numerous cylindrical heads in a large branched inflorescence; the involucral bracts are papery and whitish.
Open fields and waste areas. Late summer–fall. **G. helleri** is similar but has a sticky glandular stem. It prefers
open woods and blooms in the fall.

Mikania scandens, Climbing Hempweed (plate 512),
is one of very few twining vines among the composites,
clambering over other low vegetation in wet places. Its
leaves are opposite, triangular, and sometimes obscurely
toothed and stalked. There are numerous heads, each
containing 4 white or pale pink flowers, in small, rounded
clusters on long axillary peduncles. Summer–fall.

Ageratina altissima *(Eupatorium rugosum),* White Snakeroot (plate 513), is a common woodland species. Its
leaves are opposite, ovate with rounded bases and sharply
toothed margins, and are well-stalked. In this and the
next species there are 10 or more florets in each head.

---

*White Snakeroot has been identified as the cause of a serious disease in humans called
"milksickness," which in turn is attributed to the consumption of milk from a cow
which has contracted "trembles" by grazing on the plant. Originally reported from
North Carolina, milksickness was at times the leading cause of death from disease in
the nation; one of its victims was Nancy Hanks Lincoln, the mother of the president.
Now that the practice of keeping a family cow has been largely superseded by the
pooling of milk from multiple suppliers, the incidence of the disease is greatly reduced.*

A. aromatica (*Eupatorium aromaticum*) is similar to the last but its leaves are smaller, have blunt teeth, and are on short petioles (under ³/₄ inch).

Eupatorium spp. The *Eupatoriums*, collectively known as Thoroughworts, are medium to tall plants with opposite or whorled leaves and round- or flat-topped inflorescences. The flowers, all of which are perfect, are white in the first six species described below, purplish in the next three, and blue in the last. Typically, they bloom in late summer and fall.

E. serotinum, Late-flowering Thoroughwort, has longer, well-stalked leaves with lanceolate blades tapered at the base. Old fields and waste ground. Fall.

E. album, White Boneset, has narrow, lanceolate leaves tapered at the base. The heads have 5 white florets each, and the involucral bracts are white-tipped. Dry woods and fields.

In **E. sessilifolium,** Upland Boneset, the lanceolate leaves are rounded at the base and are finely toothed and sessile. Except in the inflorescence, it is a smooth plant.

E. rotundifolium (incl. **E. pubescens** and **E. pilosum**), Round-leaved Thoroughwort, is another 5-flowered woodland species with sessile leaves, but it is soft-hairy throughout. The leaves are ovate to nearly round, but several varieties based on differences in leaf shape have been proposed.

E. perfoliatum, known simply as Boneset (plate 514), is one of the easiest to recognize, due to the connate bases of its long-tapering, wrinkled, lanceolate leaves. There are 9 or more white flowers in each head in the flat-topped inflorescence. It is a tall plant growing in moist ground.

E. hyssopifolium, Hyssop-leaved Thoroughwort, is our only white-flowered species with verticillate leaves; they are linear and are arranged in whorls of 4, with tufts of much smaller leaves in the axils. The inflorescence is rounded and composed of numerous 5-flowered heads. Dry open places.

Our other three *Eupatoriums* with whorled leaves have purplish flowers and are popularly known as Joe-Pye Weeds. They are tall plants with toothed lanceolate leaves and large branched inflorescences. **E. fistulosum** (plate 515) has a hollow glaucous stem tinged with purple throughout. The leaves are blunt-toothed and in whorls of 4 to 7. There are many heads of between 5 and 8 pinkish purple flowers in a rounded inflorescence. Wet meadows and thickets.

In **E. purpureum** the stem is solid and generally greenish but purple at the nodes. There are only 3 or 4 leaves, with sharp teeth, in each whorl. The inflorescence is similar to the preceding except that the flowers are pale dull pink. Moist to dry woodlands. Small plants with mostly paired leaves have been segregated as **var. amoena.**

E. maculatum also has a solid stem but it is dark purple, or at least purple-spotted. The sharp-toothed leaves are in whorls of 4 or 5. The inflorescence is distinctly flat-topped and the heads 8- to 15-flowered; the corollas are purple. Moist ground.

Joe-Pye Weeds are among the few plants to bear the name of an American Indian. Joe Pye was an itinerant herb doctor who was active in New England during the latter part of the eighteenth century.

E. coelestinum (*Conoclinium coelestinum*), Mistflower (plate 516), is our only species with blue-violet flowers. Its leaves are triangular, blunt-toothed, stalked, and opposite. There are many flowers tightly packed into bell-shaped heads, which form a compact inflorescence. This attractive wildflower is frequently transplanted to gardens, where it has been called Wild or Hardy Ageratum because of its similarity to the more common annual which bears the generic name *Ageratum*. Moist thickets and woodland margins. Summer-fall.

Kuhnia eupatorioides, False Boneset. As both the scientific and common names indicate, this might be mistaken for a *Eupatorium*, but its alternate leaves furnish an instant clue. Examination will also reveal that the bristles of the pappus are plumose. The flowers are creamy white, and the heads are in small terminal clusters forming an open, branched inflorescence. Dry woods and open areas. Summer–fall

Erechtites hieracifolia, Pilewort, Fireweed (plate 517). This is a tall, coarse weed with sessile, sharply toothed lanceolate cauline leaves, and a terminal panicle. The heads are about $^1/_2$ inch high, cylindrical but swollen at the base, the principal bracts equal and in a single series. The whitish corollas are barely visible above the involucres. Old fields, burns, and waste areas. Summer–fall.

Arnoglossum atriplicifolium (*Cacalia atriplicifolia*), Pale Indian Plantain (plate 518). These are tall, smooth plants with alternate, petioled cauline leaves and flat-topped terminal clusters of whitish or greenish flowers in summer and fall. This species has palmately-veined, kidney-shaped leaves that are irregularly toothed or shallowly lobed, and glaucous beneath (as is the round stem). The involucres have 5 main bracts and are 5-flowered. Dry woods and roadsides.

A. muhlenbergii (*Cacalia muhlenbergii*), Great Indian Plantain, is similar but has a strongly grooved stem and its leaves are green beneath. It is not glaucous. Open woods.

Synosma suaveolens (*Cacalia suaveolens*), Sweet-scented Indian Plantain, has pinnately-veined, triangular-hastate leaves with fine, sharp teeth on the margins. The heads are 20- to 40-flowered, with about 12 principal bracts in the involucres. Moist woods.

Rugelia nudicaulis (*Cacalia rugelia, Senecio rugelia*), Rugel's Indian Plantain (plate 519), is very unlike the preceding three species. It is a much shorter plant and has ovate, petioled basal leaves but only small, sessile, bractlike leaves on the stem. The flower heads are $^1/_2$ inch long, bell-shaped, and horizontal or nodding. Although it is not known to occur anywhere outside the boundaries of Great Smoky Mountains National Park, it grows there in large colonies at high elevations. Summer.

Tanacetum vulgare, Tansy (plate 520), is an old-time garden herb that has escaped to roadsides and waste ground. Its strongly aromatic foliage is smooth, the leaves bipinnately dissected. The numerous disk flowers are rich yellow and bloom in summer and fall. They are in dense, flattish, button-like heads up to $^3/_8$ inch wide, the entire inflorescence more or less flat-topped.

Matricaria matricarioides, Pineapple Weed (plate 521). This is a low, branching weed of open places, with an odor reminiscent of pineapple when bruised. The leaves are once- to thrice-pin-natifid, the segments linear. Its greenish yellow heads are conical, $^1/_4$ inch wide, at the ends of the branches. The tiny disk corollas are 4-toothed. Summer-fall.

Heads discoid; flowers pink, purple or blue
(See also Eupatorium)

Cirsium spp., Thistle. All of our Thistles in this genus have spiny leaves, most of them pinnatifid and often white-hairy beneath. The flowers are magenta to purplish pink (occasionally white) with slender tubular corollas, and the pappus is plumose. They inhabit fields and other open places and bloom in summer and fall.

Cirsium vulgare (*Carduus lanceolatus*), Bull Thistle (plate 522), is our only *Cirsium* in which the leaf bases extend downward along the stem, forming prickly wings. The heads are 1 $^1/_2$ inch or more high, and the involucral bracts are armed with long, sharp spines.

C. arvense (*Carduus arvensis*), misnamed Canada Thistle (plate 523), is actually a native of Eurasia. It is a smooth-stemmed plant that spreads by underground stems to become a pervasive and undesirable weed. There are numerous small ($^1/_2$- to $^3/_4$-inch) heads of pink-purple flowers in a branched inflorescence. The involucres are smooth or at most have minute prickles on the outer bracts.

C. pumilum (*Carduus pumilus*), Pasture Thistle, can be identified by its hairy stem and by its leaves, which are green on both sides. The flower heads are 2 to 3 inches across, and the involucral bracts have slender, erect spines.

C. muticum (*Carduus muticus*), Swamp Thistle (plate 524). In this smooth-stemmed species the involucres are spineless but are covered with cobwebby hairs, and each bract has a sticky glutinous midrib.

C. discolor (*Carduus discolor*), Field Thistle. In this and the next two species the stems are smooth and the leaves white-woolly on the undersides. *C. discolor* is tall (up to 6 feet) with 1 ¹/₂- to 2-inch-high flower heads closely subtended by several narrow, ascending leaves. The principal leaves are deeply pinnatifid.

C. altissimum (*Carduus altissimus*), Tall Thistle (plate 525), is similar but most of its cauline leaves are unlobed, although they may have large teeth. The flower heads are pink.

C. carolinianum (*Carduus carolinianus*) has narrow oblanceolate leaves, mostly ¹/₂ inch or less in width, with small prickles on the margins. The flower heads are less than 3/4 inch across, with slender spines on the involucres, and the stem beneath the flowers is naked except for a few small bracts.

C. horridulum (*Carduus spinosissimus*), Yellow Thistle (plate 526), is a robust plant with large purple heads (frequently measuring 2 inches or more across) subtended by erect, narrow, spiny bracts. Occasionally found in waste areas, blooming in spring and early summer. (Yellow-flowered plants are found only on the coastal plain, where it is much more common.)

Carduus nutans, Musk or Nodding Thistle (plate 527). The plants in this genus are separated from *Cirsium* on the basis of the pappus, which consists of simple rather than feathery bristles. *C. nutans* has a 1 ¹/₂-inch-wide solitary, nodding red-purple head with broad purplish reflexed involucral bracts.

C. acanthoides, Plumeless Thistle, is similar but has smaller, erect heads in clusters and narrower involucral bracts.

Scotch Thistle (Onopordon acanthium) *is an Old World native only occasionally naturalized in America. According to legend, it was honored as the national emblem of Scotland for its part in thwarting an attempted Norse invasion. It seems that a band of enemy soldiers, hoping to gain the advantage of surprise in the darkness, removed their footgear and promptly stepped on the spiny plants. The resultant outcry alerted the sleeping Scots and enabled them to repel the attackers.*

Arctium minus, Common Burdock (plate 528). The rose-purple flowers of this bushy plant suggest those of a Thistle, but the involucres are globose, and their bracts terminate in long spines that are sharply hooked at the apex. The ³/₄-inch-wide heads are short-stalked, and the leaves, which are ovate, are neither prickly nor spiny. Open areas. Summer–fall.

Centaurea cyanus, Bachelor's Button, Cornflower (plate 529), is native to the Mediterranean region but is widely cultivated in the United States and has escaped to fields and waste places, blooming in spring and summer. Its flowers are typically deep blue but vary to pink and even white. In this genus the outer florets have large, bilaterally symmetric corollas, thus simulating ray flowers. The involucral bracts have a narrow, dark, fringed margin. Its leaves are simple and linear.

C. maculosa, Spotted Knapweed (plate 530), is much-branched and has pinnatifid foliage with linear segments. Its flowers are pink-purple, and the involucral bracts have a short, dark, fringed tip. Open areas. Summer.

C. jacea, Brown Knapweed, has wider leaves, mostly simple. The green portions of the involucral bracts are hidden beneath the overlapping appendages, which are light brown, papery, and irregularly lacerate. Fields and roadsides. Summer–fall.

Elephantopus tomentosus, Elephant's Foot. The little pale purple flowers of *Elephantopus* are discoid but asymmetrically 5-lobed and are arranged radially to form more or less circular 2- to 5-flowered heads; these in turn are gathered into glomerules which are subtended by relatively large leaflike bracts. In *E. tomentosus* the leaves are mostly basal, broadly oblong, and lie flat on the ground; those on the stem are reduced to small bracts.

In **E. carolinianus** (plate 531) the basal leaves are usually absent at flowering, but there are well-developed leaves at the bases of the branches. Open woods. Summer–fall.

Pluchea camphorata, Camphorweed (plate 532), is a strong-scented plant with toothed lanceolate to ovate leaves and a rather flat, congested inflorescence of pink flowers. It can be distinguished from the *Eupatoriums* by its alternate leaves and the fact that the outer flowers are without stamens. Wet soil. Summer–fall.

Gamochaeta purpurea (*Gnaphalium purpureum*), Purple Cudweed (plate 533), is a small plant with oblanceolate leaves. The heads are densely clustered in a narrow, often interrupted, spikelike terminal inflorescence; the bracts are reddish to brownish purple. Fields and other open areas. Spring–summer.

Liatris spp., Blazing Star. The Blazing Stars are tall, unbranched plants with mostly narrow, entire alternate leaves and spikes or racemes of rose-purple flower heads. They sometimes hybridize, and determination of species often involves the number of flowers in a head, minor characteristics of the involucral bracts, etc. Their flowering time is from late summer to fall.

Liatris graminifolia, Grass-leaved Blazing Star, is one of a confusing group of narrow-leaved species. It has numerous leaves less than $1/4$ inch wide. The heads average about 10 flowers each and are almost sessile; the involucral bracts are blunt at the apex. Dry woods and fields.

Similar species include **L. microcephala,** which usually has only 5-flowered heads on slender peduncles; **L. regimontis,** with fewer leaves and short-stalked heads with 6 to 12 flowers and glandular-dotted, sharp-pointed bracts; and **L. turgida,** in which the lowermost leaves are oblanceolate and from $1/4$ to 1 inch wide, and the heads may have as many as 20 flowers. All of these are usually found in fields and open woods.

L. helleri, a rare plant of rocky wooded habitats, resembles L. *turgida* but differs in that its pappus is short— no more than half as long as the corolla tube.

L. spicata, Dense Blazing Star (plate 534), also has narrow leaves. Its heads are sessile and only 8-flowered, but very numerous and crowded. The involucres are slender, and the bracts few, blunt, and often purplish. Moist habitats.

Yet another narrow-leaved species is **L. squarrosa,** known as Scaly Blazing Star. It has more than 20 flowers per head, and its bracts are lanceolate with long-pointed tips, and spreading. It is our only species with plumose rather than barbellate pappus. Dry, open sites.

In **L. scariosa** the lowest leaves are on long petioles with blades up to 1 ½ inch wide, the upper ones much smaller. The heads are many-flowered, and the bracts are loose and broadly rounded, with purple or white ciliate borders. Dry woodlands.

L. aspera, Rough Blazing Star (plate 535), also has rounded bracts, but they have irregularly toothed pink or whitish margins. Rocky woods.

Vernonia spp., Ironweed. The Ironweeds are tall plants with the general aspect of *Eupatoriums* but with narrower, alternate leaves, more open inflorescences, and darker purple flowers. The pappus is double, consisting of long inner and short outer bristles. All bloom in late summer or early fall.

Vernonia noveboracensis, New York Ironweed (plate 536), has flowers of an intense deep purple-violet color which extends in some degree to the herbage and even the pappus. The heads average about 30 flowers each and are in a flat-topped inflorescence. The involucral bracts are abruptly narrowed to a hairlike tip. Moist woods and open areas.

V. glauca also has many-flowered heads, in an irregular inflorescence. The apexes of the bracts are slightly shorter than in the preceding species, and the pappus is straw-colored. The leaf undersides are glaucous and distinctly paler than the upper surface. Upland woods.

V. gigantea *(V. altissima)*, Tall Ironweed, has numerous heads of fewer than 30 flowers each. The bracts are rounded, short-pointed, and purplish, and the pappus is purple. Moist habitats.

V. flaccidifolia is a glaucous, few-flowered species, with blunt green bracts with purple edges and tawny pappus. Woods.

Marshallia obovata, Barbara's Buttons (plate 537), has a few entire, narrowly oblanceolate leaves in a basal rosette; the cauline leaves are slightly smaller and are confined to the lower half of the stem. The tubular florets are white with conspicuous lobes, in a solitary 1 $1/4$-inch head terminating the naked portion of the stem. The involucral bracts are obtuse. Old fields and open woods. Spring.

Heads radiate; rays yellow

Rudbeckia hirta, Black-eyed Susan (plate 538). Our most common species, this is a hairy plant with alternate, entire or toothed (but not lobed) leaves. The receptacle is dark purple or brown, $1/2$ to $3/4$ inch wide and decidedly ovoid or hemispheric, the disk flowers fertile. There are 10 or more sterile rays, yellow sometimes shading to orange, up to 1 $1/2$ inches long.

R. fulgida, Orange Coneflower, is a similar but variable species, frequently displaying more orange in its ray flowers. One way to distinguish between the two is to examine the **Y**-shaped styles of the disk flowers; the prongs are slender and pointed in R. *hirta,* short and blunt in R. *fulgida.* Both are found in a variety of habitats, blooming in summer and fall.

R. triloba, Thin-leaved Coneflower, is hairy, and its flower heads resemble those of the preceding but tend to be more numerous and smaller, and the rays fewer but deeper in color. The principal difference, however, is that at least some of the lower leaves of R. *triloba* are deeply lobed into 3 or more segments. Moist, open woods. Summer-fall.

R. laciniata, Green-headed Coneflower (plate 539), is tall (5 feet or more in height) and smooth, with many of its leaves deeply ternately or pinnately lobed. The

disk is strongly rounded, up to $^3/_4$ inch across, and grayish or yellowish green. The yellow rays are 1 $^1/_2$ inches
long, often drooping. Woodland edges and meadows,
in moist soil. Summer.

Ratibida pinnata, Gray-headed Coneflower, is characterized by a columnar central disk, which in this species
is gray-brown and higher ($^1/_2$ to $^3/_4$ inch) than wide.
There are 5 to 10 long, strongly reflexed yellow rays.
The leaves are alternate and pinnatifid, the segments
lanceolate. Dry woods and fields. Summer.

Helenium autumnale (*H. virginicum*), Sneezeweed
(plate 540). The flowers in this genus are remarkable
for their large, many-flowered, nearly globose disks and
their wedge-shaped, 3-lobed yellow rays, which are
slightly deflexed. In *H. autumnale* the disk is yellow and
the ray flowers pistillate; the heads are about 1 $^1/_4$ inch
wide. Its numerous leaves are narrowly ovate and mostly
toothed, their bases decurrent on the stem. Moist, low
ground. Late summer–fall.

H. flexuosum (*H. nudiflorum*), Purple-headed Sneezeweed (plate 541), has fewer leaves, and they are narrower and entire; the stem is broadly winged. The flowers
are similar to the above but the disk is dark purple or
purplish brown, and the ray flowers are neutral. Moist
fields and waste places. Summer–fall.

H. amarum, Bitterweed (plate 542), is a low, much-
branched plant with very numerous linear, even threadlike, leaves. There are many flower heads, with yellow
disks about $^3/_8$ inch wide and 5 to 10 rays. Fields, roadsides, and waste areas. Summer–fall.

---

*The name Sneezeweed reflects the practice of inhaling the dried leaves as a snuff to
induce sneezing and thus clear the nasal passages to relieve cold symptoms. American colonists found a very different use for the plants: the preparation of a "fish
poison," which was introduced into streams to temporarily stun fish and cause
them to float to the surface.*

Helianthus spp., Sunflower. Its many similar, often hy-
bridizing, species make this genus one of the more frus-
trating groups of composites. Only a few characteristics
are consistent: The receptacle is more or less flat, the

disk (except in the first three species) is yellow, the ray flowers are neutral, and they bloom in summer and fall.

Helianthus annuus, Common Sunflower (plate 543), is the best-known, although many who are familiar with the enormous cultivated plants may fail to recognize it in its much smaller native form. It can be identified by its rough-hairy leaves, which are ovate to triangular with cordate bases and have long stalks. The disk is reddish brown. Open places.

H. angustifolius, Narrow-leaved Sunflower, has a scabrous stem and stiff, narrow (under $^1/_2$-inch) leaves, alternate and sessile on the stem. The heads have a $^3/_8$-inch-wide dark purple disk, and rays about 1 inch long. Moist locations.

In *H. atrorubens,* Hairy Wood Sunflower (plate 544), the stem is hairy but the leaves are mostly opposite. Almost all are near the base of the plant and have ovate blades and broadly winged petioles; the few upper ones are greatly reduced in size. The flower heads, which are borne at the ends of long naked branches, are similar to those of the preceding species but have a reddish disk. Open woods.

In *H. occidentalis* the stem is virtually smooth above. Most of its leaves are ovate and near the base (the others are few and very small), but their blades are abruptly contracted to a long, slender petiole. The disk is yellow. Plants with their middle leaves not greatly reduced in size, and the lower ones often absent at flowering, have been segregated as **var.** *dowellianus.* Dry soil.

H. mollis is densely covered with whitish hairs throughout. Its leaves are mostly opposite and sessile, broadly lanceolate or ovate with rounded clasping bases. Its showy heads have 15 to 30 ray flowers 1 inch long. Dry soil.

H. divaricatus, Woodland Sunflower, also has opposite and virtually sessile leaves, but they taper gradually to a long tip and have abruptly rounded bases that do not clasp. The stem is essentially smooth. There are less than 15 rays, about 1 inch long. Dry woods and roadsides.

The leaves of another smooth plant, **H. microcephalus,** Small Wood Sunflower (plate 545), resemble those of the preceding but they are on petioles about $1/4$ inch long. The heads are only 1 to 1 $1/2$ inches across, with 5 to 8 rays. Woodlands.

H. hirsutus, Rough Sunflower, is bristly-hairy, and its leaves are similar to those of H. *divaricatus* but are very rough on the upper surface and short-petioled. The flower heads are much larger than in H. *microcephalus,* however, and have about 12 rays. Dry woods and roadsides.

H. strumosus, Pale-leaved Sunflower, is one of several species with leaves that taper to a winged petiole. The lanceolate blades are thick, whitish beneath, and slightly toothed or entire. The stem is smooth and glaucous below the inflorescence. The disk is small compared to the 8 to 15 rays, which may be 1 $1/2$ inches long. Woodlands.

H. decapetalus, Thin-leaved Sunflower, is another smooth-stemmed species. Its opposite leaves are thin, ovate, green beneath, and conspicuously toothed, on winged petioles. Its flowers are similar to those of H. *strumosus.* Woodlands.

H. tuberosus, Jerusalem Artichoke (plate 546), is a tall, hairy plant. Its leaves, which are mostly alternate, are thick and rough-hairy, the ovate blades coarsely toothed and tapering to a long, winged petiole. The flower heads have up to 20 rays 1 $1/2$ inches long but a disk less than $1/2$ inch wide. The involucral bracts have spreading tips. Widespread, in woodland borders and waste ground.

The first part of the name "Jerusalem Artichoke" has nothing to do with the ancient city but is a corruption of girasole, *an Italian word that simply means "sunflower." This species was cultivated for its crisp, edible tubers before the first Europeans reached the Americas; they are marketed today as "sun chokes."*

In the strictly southern **H. resinosus** (H. *tomentosus),* Gray Sunflower (plate 547), the lanceolate-ovate leaves are mostly alternate and scabrous above but velvety beneath, and the petioles winged; the stem is hairy. The flowers have about 12 wide, overlapping, light yellow

1 ¹/₂-inch rays, a disk up to 1 inch across, and spreading to reflexed bracts. Woods and meadows.

H. glaucophyllus is a smooth species with opposite lanceolate leaves that are scabrous above and glaucous beneath. The petioles are up to 1 ¹/₂ inches long, not winged. The rays are 1 inch or longer. Moist woods and roadsides.

H. giganteus, Tall Sunflower, has a hairy reddish stem and mostly alternate, narrowly lanceolate leaves, green on both sides and tapering to a short petiole. Although this species may grow to 10 feet in height, its flowers are proportionately small, often less than 2 inches in diameter, with 10 to 20 rays, and narrow, spreading bracts. Wet places.

H. laevigatus is essentially smooth throughout. Its leaves are similar to those of the preceding, but the flower heads are about 1 ¹/₂ inches wide and have only 5 to 10 short rays. Woods.

Heliopsis helianthoides, Ox-eye, False Sunflower (plate 548). The differences between this and the true Sunflowers (*Helianthus*) are mostly technical; probably the most easily observed clue is the 2-pronged style of the rays, indicating fertile rather than neutral flowers. The leaves are ovate, toothed, long-stalked, and all opposite. The attractive flower heads are about 2 inches wide, with a yellow conical disk and about 12 rays. Woods and thickets. Summer.

Tetragonotheca helianthoides, Pineland Ginseng (plate 549), is noteworthy for its square involucre, formed by 4 broad, ovate bracts which may be 1 inch long. It is a stout plant with wide, opposite, sessile leaves with coarse teeth. The heads are solitary, with greenish disk flowers and 6 to 10 rays up to 1 ¹/₂ inches long and ³/₈ inch wide, 3-lobed at the apex, and abruptly narrowed at the base. Dry, open locations. Spring–summer.

Coreopsis spp., Tickseed. Our plants in this genus have opposite leaves and showy flowers; both disk and ray florets are completely yellow except in the first two species. The involucral bracts are in 2 series, the outer green and leaflike, and spreading. The pappus, if present, is of

2 teeth or awns which may be barbed but never retrorsely. They occur in a variety of open habitats.

Coreopsis tinctoria, Calliopsis (plate 550), is a western species frequently escaping from cultivation here and blooming in summer and fall. It is distinguished by its reddish purple disk and by a large red-brown blotch at the base of the yellow rays, which are conspicuously toothed. The heads are numerous and measure about 1 inch across. The leaves are petioled, and most are once or twice pinnately divided into very narrow linear segments.

C. tripteris, Tall Coreopsis, also has a purplish disk, but the rays are wholly yellow and without teeth at the apex. The principal leaves are palmately divided into 3 narrow lanceolate segments, and most are petioled. Summer–fall.

C. major, Greater Tickseed (plate 551), also has ternately divided leaves but they are sessile, so that each pair simulates a whorl of 6 leaves. Both the disk and ray florets are yellow; the latter may or may not be toothed. Summer.

In **C. verticillata,** misnamed Whorled Coreopsis, the leaves again are sessile but not whorled; each is ternately dissected (and often pinnately redivided) into extremely narrow, threadlike segments. The rays are up to 1 inch long and untoothed. Summer.

C. pubescens, Star Tickseed, is a leafy-stemmed species native to the South. Its leaves are elliptic to ovate, under 5 inches long, entire or with 1 or 2 pairs of similar but smaller basal lobes, generally hairy. The rays are broad and bear several teeth. Summer–fall.

C. lanceolata, Lance-leaved Coreopsis, is very widely cultivated and therefore much better known than the preceding species, from which it differs in being smooth and having narrower and longer (up to 8-inch) leaves; most arise from near the base of the plant. Summer.

C. grandiflora is similar to the last, but its leaves, which continue upward on the stem, are pinnately divided with lateral lobes less than $1/4$ inch wide. Summer.

C. auriculata, Eared Coreopsis (plate 552), is stoloniferous and has mostly basal leaves; they are petioled, ovate to nearly round, and often have a pair of earlike lobes at the base of the blade. The heads are long-stalked with broad, orange-yellow rays bearing 4 prominent teeth. Woodlands. Spring–summer.

C. latifolia is an uncommon woodland species with ovate, unlobed, short-stalked leaves that are coarsely toothed on their margins—a unique characteristic. Its rays are less than $^3/_4$ inch long. Summer.

Bidens spp., Beggar Ticks. Plants in this genus share a number of characters with *Coreopsis*—among them opposite leaves, involucral bracts in two dissimilar series, and pappus consisting of awns. Most species have yellow rays, but these range from large and showy to small and inconspicuous, and sometimes are absent altogether. More often than not, these plants are noticed less for their flowers than for the achenes, which are surmounted by stiff awns that, unlike those of *Coreopsis*, are retrorsely barbed and attach to one's clothing at the slightest touch (see *B. tripartita*, plate 555). Our species are usually found in wet fields, roadside ditches, and waste areas, and flower in summer and fall.

Bidens polylepis, Bur Marigold (plate 553 and plate 554), is a handsome, moisture-loving species admiringly referred to as Ditch Daisy. Its leaves are pinnately or occasionally bipinnately dissected into narrow, coarsely toothed segments. The bright yellow rays are 1 inch or more long and surround a $^1/_2$-inch disk. There are 12 or more outer involucral bracts, which are ciliate and longer than the inner ones. The pappus is of 2 short awns or none.

B. aristosa is a similar species found less frequently in the mountains. Its outer bracts are 10 or fewer, shorter than the inner, and have smooth margins.

B. cernua, Nodding Bur Marigold, has simple lanceolate leaves, sessile and sometimes connate at the base, usually toothed. The flower heads droop with age; they have 6 to 8 rays, which are less than $^1/_2$ inch long (or they may be absent) and often exceeded by the narrow outer bracts. The pappus is of 4 barbed awns.

B. bipinnata, Spanish Needles, is our only species with consistently bipinnately compound leaves. The heads are erect and very narrow; the achenes bear 3 or 4 awns and are divergent at maturity. There are commonly 1 to 3 very short rays, sometimes none.

In **B. tripartita,** Beggar Ticks (plate 555), the leaves may be simple or deeply 3-lobed, the segments lanceolate and toothed; most have conspicuously winged petioles. The heads are usually discoid but short rays are sometimes present. The outer bracts are long and leaflike. There are 2 to 4 awns on each achene.

The leaves of **B. frondosa** (plate 556) are pinnately compound with 3 or 5 toothed lanceolate segments. The heads are orange, usually without rays. There are about 8 leafy outer bracts, which greatly exceed the disk. The pappus is blackish and 2-awned. **B. vulgata** is similar but has a yellow disk and about 13 unequal outer bracts.

Pityopsis graminifolia (*Chrysopsis graminifolia*), Grass-leaved Golden Aster (plate 557). In this species the stem and the alternate linear leaves, which are parallel-veined, appear silvery from their covering of long white appressed hairs; the basal leaves are much longer than those above. Both the disk flowers and the $1/2$-inch rays (about 12) are fertile. The pappus is double, the inner of hairy bristles, the outer of small scales. Dry, sandy places. Summer–fall.

P. ruthii (*Chrysopsis ruthii, Heterotheca ruthii*), Ruth's Golden Aster, is a rare species growing in rocks along river gorges and blooming in late summer or fall. It is a low, much-branched plant with narrow silvery leaves (basal shorter than cauline), glandular in the inflorescence. The heads are small, with about 12 rays $3/8$ inch long.

Chrysopsis mariana (*Heterotheca mariana*), Maryland Golden Aster (plate 558), has alternate lanceolate, finely-toothed leaves, the lower ones larger and pinnately-veined, not silver-hairy at maturity. Its flower heads are similar to those of *Pityopsis graminifolia* except for having more and slightly longer rays; the involucres and flower stalks are sticky-glandular. Open woods and fields. Summer–fall.

Heterotheca subaxillaris, Camphorweed (plate 559), resembles plants in the genus *Chrysopsis* except for the fact that its ray flowers have no pappus. It is a hairy, glandular plant with firm, wavy-edged ovate leaves, the basal ones often absent, the cauline ones sessile or clasping. Sandy fields, roadsides, and wood margins. Summer–fall.

In hot weather the leaves of Camphorweed twist so that their edges face the sun, thereby retarding evaporation by reducing the amount of exposed surface area.

Senecio spp., Groundsel, Ragwort. These are plants with both basal and alternate cauline leaves, the latter sometimes very different in shape from the former. The involucral bracts are essentially equal and are subtended by a ring of much smaller ones referred to as bracteoles. The ray flowers (absent in the first species) and the disk flowers are fertile; the pappus is of numerous white bristles. Most bloom in spring or summer.

Senecio vulgaris, Common Groundsel, is our only species without rays (but see *Rugelia nudicaulis,* which was formerly in this genus). There are numerous basal and cauline oblanceolate leaves, all coarsely and irregularly pinnatifid. The flower heads are cylindrical, $^3/_8$ inch long, at the ends of branches; the bracteoles have black tips, and the pappus is longer than the florets. Waste ground. Spring–fall.

S. millefolium, a rare endemic of dry rock outcrops, has numerous flower heads. It has mostly basal leaves, which are twice or more pinnately dissected, the segments linear and sharp-pointed; it is distinguished from our other species by the fact that none of the leaf divisions is as much as $^1/_8$ inch wide. Spring–early summer.

S. anonymus (*S. smallii*), Squaw Weed, Small's Ragwort (plate 560). In all of our remaining species the basal leaves are petioled, and the cauline leaves pinnatifid or toothed and at least the uppermost ones sessile. *S. anonymous* is a southern species. It has elongate basal leaves (frequently 6 inches or longer overall) with elliptic, blunt-tipped blades toothed or scalloped, some pinnately lobed, and tapering gradually to a long petiole. The few cauline leaves are smaller and deeply pinnatifid, becoming sessile above. The flower heads are 1/2 inch across and very numerous. Open woods, fields, and roadsides.

S. pauperculus, Balsam Ragwort, is a plant of more northern distribution but may be found in our area in wet meadows, blooming in spring. Its basal leaves are smaller (under 4 inches), the blades elliptic or oblanceolate, toothed, rounded at the summit, and taper more abruptly at the base to a long petiole. The cauline leaves are pinnatifid or irregularly incised. There are generally fewer than 20 heads.

S. obovatus, Roundleaf Ragwort, differs from the last two species in having well-developed stolons. Its basal leaves have obovate or nearly round blades tapered at the base. Wooded, rocky slopes, especially in calcareous soil.

S. tomentosus, aptly named Woolly Ragwort (plate 561), has conspicuously white-woolly stems, leaf undersides, and inflorescences. All of the leaves, basal and cauline, are lanceolate to ovate with scalloped or toothed margins (not pinnatifid). Granite outcrops and other dry, open sites.

S. aureus, Golden Ragwort (plate 562). In this and the next two species the blades of the basal leaves are widest just above their bases, which are cordate or truncate. S. aureus has rounded ovate leaves, with a distinctly cordate base and blunt-toothed margins, on long petioles; the undersides, as well as the involucres, are often dark purple. The cauline leaves are greatly reduced in size and are narrow, sessile, and pinnatifid. The heads are $^3/_4$ to 1 inch wide, with an orange disk and yellow rays. Woods and fields.

In **S. schweinitzianus** (S. robbinsii), the basal leaves are long-petioled, but the blades are lanceolate-ovate and more than twice as long as wide, toothed on the margins, with a sharp-pointed apex and a truncate base. The cauline leaves are as in S. aureus. This is a disjunct northern species found on mountain balds in the South.

S. *plattensis* has smaller basal leaves on shorter petioles; the blades are broadly elliptic to ovate and are abruptly contracted at the base. The cauline leaves are smaller and somewhat pinnatifid. Like *S. tomentosus*, the plant remains white with woolly hairs until after flowering. Dry, open places.

Chrysogonum *virginianum*, Green-and-Gold (plate 563), has few but attractive flowers, each with 5 broad, fertile yellow rays up to $^1/_2$ inch long; the disk is about $^1/_4$ inch across, its florets sterile. It begins to bloom early in the spring before the foliage attains full size, continuing into summer. The hairy leaves are opposite, long-petioled, with ovate, scalloped blades. Woodlands.

Silphium *perfoliatum*, Cup Plant. The *Silphiums* are tall, coarse plants with heads of fertile ray flowers and staminate disk flowers in a widely branched inflorescence. This species is named for its leaves, which are connate. The stem is 4-angled and smooth, but the foliage is rough. Its heads are 2 to 3 inches across, with 20 or more narrow rays. Moist woods and thickets. Summer.

S. *compositum*, Rosinweed (plate 564). Unlike our other species, this has mostly basal leaves; they are large, about as broad as long, kidney-shaped, and deeply lobed or coarsely toothed. The heads have a 3/8-inch disk and 5 to 10 rays less than $^3/_4$ inch long. Dry woods and waste places. Spring–fall.

S. *trifoliatum*, Whorled Rosinweed, is unique among our species in having many of its leaves in whorls of 3 or 4; they are rough, lanceolate, short-stalked, and more or less toothed. The stem is smooth and often glaucous. The heads are about 2 inches wide. Woods and fields. Summer.

In **S. *dentatum*** the leaves are narrow and may be either alternate or paired and usually are toothed. The stem is essentially smooth. There are about 20 rays, 1 inch long. **S. *asteriscus*** differs in having a coarsely hairy stem and only about 10 rays per head. Its leaves are mostly opposite. Dry woods and fields. Summer.

Verbesina occidentalis, Crown-beard (plate 565). Our plants in this genus are tall, with more or less winged stems and simple leaves. The disk flowers are perfect, with divided styles, and the pappus is of 2 awns. They bloom in late summer and fall. *V. occidentalis* has opposite ovate leaves. The flower heads have 5 or fewer irregularly spaced, drooping yellow rays between $^1/_4$ and $^3/_4$ inch long. The involucral bracts are erect. Woods and open places.

V. alternifolia (*Actinomeris alternifolia*), Wingstem, can be identified by its alternate lanceolate leaves and by the involucral bracts, which are reflexed. Its flowers have 2 to 10 drooping yellow rays up to 1 inch long. Moist locations.

V. virginica, Tickweed (plate 566), also has alternate leaves, but its rays are white, less than $^3/_8$ inch long, and no more than 5 per head; the disk flowers also are white. Woods and open spaces.

Smallanthus uvedalia (*Polymnia uvedalia*), Bearsfoot, Yellow Leafcup (plate 567). The Leafcups are tall plants with large opposite leaves. The disk flowers are yellow and sterile, with undivided styles; there is no pappus. *S. uvedalia* has broad leaves that are palmately but irregularly lobed with winged petioles, or sessile above. There are from 8 to 15 yellow rays $^1/_2$ to $^3/_4$ inch long. Woods and meadows. Summer–fall.

Inula helenium, Elecampane. This robust European introduction, which has become naturalized in fields and along roadsides, is unmistakable. Its flowers have very numerous, slender 1-inch yellow rays in a single series, surrounding an even wider yellow disk. The outer bracts of the involucre are large and leaflike. The large elliptic leaves are irregularly toothed and woolly beneath, the upper ones clasping the stem. Summer.

Tussilago farfara, Coltsfoot (plate 568). The solitary flowers of Coltsfoot arise in early spring from a creeping rhizome on short, scaly-bracted scapes. (The leaves, which develop later, are basal, nearly round, and shallowly lobed with a cordate base.) The 1-inch heads superficially resemble those of Dandelions but have very numerous narrow, yellow fertile rays in several overlapping series and similarly colored but sterile disk florets. Waste ground.

The history of treating coughs with Coltsfoot goes all the way back to the ancient Greeks, who boiled the leaves with honey to make a smooth, pleasant-tasting syrup. This practice is reflected in the name of the genus, which is from tussis, *the Latin word for cough.*

Solidago spp., Goldenrod. Without doubt, the Goldenrods are the most difficult genus in which to identify species—not even excepting the Asters, which are more numerous. Aside from the fact that a number of natural hybrids are known to occur, the principal reasons for this difficulty are that the flower heads are small and the rays very short, and both the ray and disk florets—with but a single exception—are yellow. Consequently, differences in floral characteristics tend to be slight and hard to discern, and this makes it more important to give attention to the vegetative parts, especially the basal leaves where they are still in evidence at anthesis. The principal flowering period for Goldenrods is fall, although about half of our species may commence blooming during the late summer months.

In the descriptions that follow, the Goldenrods are arranged roughly according to their types of inflorescence. The first few have their flower heads disposed in small, short-stalked clusters in the well-spaced axils of the upper leaves. These are followed by several in which the heads are crowded more closely together to form a narrow wandlike or clublike terminal inflorescence. The next group is a large one, consisting of species in which the branches of the panicle are long and more or less recurved, and appear secund (that is, with flower heads only on the upper side). Finally, there are three with flat-topped, corymbiform inflorescences.

Solidago bicolor, Silverrod (plate 569), can be separated from all the others at the outset by noting that it is the only species with white rays. It has a stiff, wandlike stem and a narrow, elongate panicle. It is found in dry woods and open places, often in large numbers.

S. curtisii, Curtis' Goldenrod. This and the next two species have their flowers in nearly sessile clusters in the upper leaf axils. All of these leaves, which have toothed margins, project well beyond the flower heads. There are only 3 or 4 ray florets in each head. In *S. curtisii* the stem is erect and is striate-angled and grooved; it is not glaucous. Its leaves are elliptic to lanceolate, at least 3 times as long as wide, and taper to a virtually sessile base. Rich mountain woods and roadsides.

S. caesia, Blue-stem or Wreath Goldenrod (plate 570), is similar, but its stem is arching, round in cross-section, and not grooved, and is glaucous. Moist woods.

S. flexicaulis, Broad-leaved Goldenrod, has an angled, conspicuously zigzag stem and widely spaced leaves. Its leaf blades are ovate, seldom more than twice as long as wide, and are abruptly contracted to a distinct, winged petiole. Woodlands.

S. glomerata, Skunk Goldenrod (plate 571). The next seven species, presented in no particular order, are erect plants with narrow terminal inflorescences in which some of the lower branches may become slightly elongate and ascending, but not recurved or secund. *S. glomerata* is unique in that its flower heads are exceptionally large and showy, measuring up to $^5/_8$ inch across; they are in short-stalked, few-flowered axillary clusters. It is a smooth plant, yellowish green in general appearance, with broadly elliptic, sharply toothed basal leaves up to 9 inches long, tapering to long petioles. Mountain balds and wet places at high elevations.

S. squarrosa, Stout Goldenrod, is another with a readily recognizable characteristic; in this case it is the involucral bracts, which have spreading, recurved tips. It is a coarse, smooth species with numerous large, elliptic, long-stalked, toothed leaves. The narrow terminal panicle has 10 or more rays in each head. Dry woods. Late summer–early fall.

S. roanensis, Mountain Goldenrod (plate 572), is mostly smooth. The basal and lowest cauline leaves are elliptic-obovate and long-pointed and have coarsely toothed margins; the upper ones are sessile, lanceolate, and often entire, surpassing the inflorescence—which is narrow, densely flowered, and cylindrical—only in the lower portion. There are about 10 rays, and the involucral bracts are sharp-pointed. Woodlands and roadsides.

S. erecta, Slender Goldenrod, is a smooth plant with a narrow, often interrupted, terminal inflorescence. The lower leaves are elliptic, shallowly toothed or entire, and blunt at the apex, the upper ones becoming much smaller, extending only into the lower part of the inflorescence. There are about 10 rays; the involucral bracts are rounded. Dry woods.

S. hispida, Hairy Goldenrod, has a configuration like *S. erecta*, but its stem and leaves are covered with spreading hairs, the leaf blades are more ovate and have blunt teeth, and the 7 to 14 rays are deeper yellow. Rocky, open places.

S. puberula, Downy Goldenrod, is minutely hairy throughout. Its lower leaves are oblanceolate, toothed, and petioled, the upper ones sessile and entire. The inflorescence is a narrow terminal panicle. The rays are numerous and showy. Open places.

S. petiolaris has an erect, hairy stem. Its leaves, all cauline at anthesis, are elliptic and generally sessile and entire. They are closely spaced and do not exceed their axillary flower clusters. There are 5 to 8 rays. Woods and road banks.

S. speciosa, Showy Goldenrod. The basal leaves are broadly elliptic and are entire or have only low teeth. The inflorescence is densely flowered, with numerous ascending branches (not secund), creating a pyramidal panicle. Each head has 5 large rays. Open fields and woodland borders.

S. uliginosa, Bog Goldenrod, is a smooth species with narrowly elliptic, finely toothed basal leaves tapering to a long petiole with a sheathing base. The inflorescence has a number of short, ascending or recurved-secund branches, but is much longer than broad. Wet places.

S. nemoralis, Gray Goldenrod (plate 573), is covered with fine hairs throughout. Its basal leaves are oblanceolate and stalked, with some blunt teeth; the upper ones are progressively reduced, with axillary tufts of smaller leaves. The inflorescence is pyramidal, nodding at the summit, with recurved secund branches. Old fields and waste ground.

S. *sphacelata*, False Goldenrod, is our only species with cordate basal leaves; they have toothed blades and long petioles. The panicles are longer than broad, with a few widely spreading branches, the flower heads secund. Rocky woods and open places. Late summer–early fall.

S. *ulmifolia*, Elm-leaved Goldenrod. The basal leaves of this species are usually absent at flowering; the upper cauline ones are ovate to elliptic, toothed, and sessile. The inflorescence is broad and very open, with a few long, arching, secund branches. There are only a few tiny rays in each head. Woodlands.

S. *patula*, Rough-leaved Goldenrod, has among the largest basal leaves—1 foot or more long—with toothed elliptic blades and a broad sheathing petiole. The stem is prominently ridged, and the foliage thick and strongly scabrous above. The inflorescence is an open panicle with widely spreading secund branches. The flower heads are 6- to 12-rayed. Swampy woods and wet meadows.

In **S. *arguta*,** Sharp-leaved Goldenrod (plate 574), the basal leaves are also large but are ovate, thin, and smooth above; the double-toothed blades taper abruptly to a long petiole and to a long point at the apex. The stem is often dark purple. The panicle is about as broad as long, its branches bearing recurved, densely flowered secund heads. There are 5 to 8 rays. Open woods and meadows. Midsummer–fall.

S. *juncea*, Early Goldenrod, is a smooth plant with persistent, narrowly elliptic, toothed basal leaves tapering to a long petiole. The cauline leaves are progressively smaller, becoming sessile, the upper ones with tufts of much smaller leaves in their axils. The flower heads have 7 to 13 rays. Woodland borders and fields. Summer–fall.

The basal leaves of **S. *odora*,** Sweet Goldenrod (plate 575), are usually deciduous before flowering time, which may occur as early as midsummer. The others (which may have small leaves in their axils) are lanceolate-linear, entire and sessile, and have only one principal vein. They may be further identified by their minute translucent dots and the odor of anise. The arching branches of the panicle are densely flowered; there are 3 to 5 rays. Dry, open woods.

S. rugosa, Rough-stemmed Goldenrod, as well as the next two species, has only cauline leaves when in flower; in this one they are lanceolate to ovate-lanceolate on very short petioles, toothed, with deeply impressed pinnate veins. The inflorescence is a wide-based panicle. There are 6 to 10 rays. Woods and old fields.

S. gigantea, Late Goldenrod, has a smooth stem, sometimes reddish, and usually glaucous below. The leaves are sessile, sharply toothed, and narrowly lanceolate, with two large veins running parallel to the midvein. The panicle has numerous spreading floriferous branches; the rays are 10 to 15 and showy. Open places. Despite the common name it is among the earliest to bloom.

S. canadensis (*S. altissima*), Canada or Tall Goldenrod (plate 576), is similar in many respects to *S. gigantea*, including triple-nerved leaves, but they sometimes have few teeth or none. Its stem is usually downy, never glaucous. Fields and roadsides.

S. rigida, Hard-leaved Goldenrod. Only two of our species have a flat-topped corymbiform inflorescence, and both are uncommon. *S. rigida* is a large plant with scabrous elliptic leaves, the basal ones long-petioled and with low blunt teeth, the upper ones much smaller, rounded at the base, entire and sessile. The inflorescence is showy, with large many-flowered heads on widely spreading branches. In our area it is found in widely scattered, dry, open locations.

S. spithamaea, Blue Ridge Goldenrod, is a rare endemic of the southern mountain balds, blooming in late summer. It is much smaller than the preceding (less than 1 1/2 feet high), and has smooth, sharply toothed leaves. The inflorescence is small and compact.

A third flat-topped Goldenrod, formerly named *Solidago graminifolia*, has been removed from that genus and is now known as **Euthamia graminifolia.** It is distinguished by a unique combination of characters: a corymbiform inflorescence, heads consisting of 20 or more flowers, and linear, glandular-dotted leaves.

The Goldenrods have long been blamed for causing allergies. This is an accusation that is easily understood, since they are so prevalent during the "hay fever season," but is undeserved because their pollen is heavy and sticky—and gathered by bees and butterflies—not fine and spread by the wind as is the case with the true culprits such as Ragweed.

Heads radiate; rays not yellow

(See also Solidago and Verbesina)

Echinacea purpurea, Purple Coneflower (plate 577). The members of this genus have large flower heads with more or less drooping purplish pink rays from 1 to 3 inches long, and a dome-shaped disk about 1 inch across. The latter is made bristly by the protruding tips of the receptacular bracts. This species has alternate, broadly lanceolate to ovate leaves that contract abruptly to a petiole, are rough above, and usually bear coarse teeth. Woods and fields. Summer–fall.

E. laevigata is similar but has smooth, glaucous leaves. It shares the same habitats but flowers from late spring to midsummer.

In **E. pallida** the principal leaves are basal; they are rough-hairy, narrow, tapering gradually at the base, and are untoothed. Dry, open places. Summer.

Like the Brown-eyed Susans, Purple Coneflowers are essentially prairie plants that migrated eastward with the settling of the American West, and are becoming increasingly familiar as a component of highway beautification projects. Pawnees and other tribes used the roots extensively for medicinal purposes, especially in treating rattlesnake bites. Modern medicine, which recognizes Echinacea *as the source of a versatile antibiotic, is now concentrating on it in the search for a possible cancer treatment.*

In **Polymnia canadensis,** Small-flowered Leafcup, the leaves have a few large pinnate lobes with toothed margins, and the petioles are not winged. There are 5 or fewer rays (sometimes none) which are white and less than 1/2 inch long; the disk flowers are yellow. Moist woods. Summer–fall.

Leucanthemum vulgare (*Chrysanthemum leucanthemum*), Ox-eye Daisy (plate 578). One of our most familiar wildflowers, this is the only Old World representative of this group of composites to have become so widely naturalized here. The flower heads are solitary, up to 2 inches across, with 15 to 30 white rays and a flat, bright yellow disk. The leaves are narrowly oblanceolate in outline, coarsely and irregularly lobed. Spring–late summer.

Anthemis cotula, Stinking Mayweed, Dog Fennel (plate 579). Our species of *Anthemis* resemble the Ox-eye Daisy but have somewhat smaller flowers and finely pinnately dissected foliage. They are weeds of fields, roadsides, and waste areas, blooming from late spring to fall. *A. cotula* has a fetid odor and leaves that are twice or thrice pinnatifid, with threadlike segments. The yellow disk is conical, with perfect yellow flowers; there are 10 to 20 sterile white rays up to 3/8 inch long, the entire head usually measuring 1 inch across.

A. arvensis, Field Chamomile, is odorless and its leaves have less finely cut segments. The disk is flat at first, not becoming conical until in fruit. The rays are fertile and slightly longer than in the preceding species, the heads 1 to 1 1/2 inches wide.

Bellis perennis, known as English Daisy (plate 580), sometimes occurs as a weed in lawns, flowering in spring or summer. It is a low plant with leaves concentrated near the base, usually obovate and somewhat toothed, seldom more than 1 inch long. The heads are solitary and about 3/4 inch across. The disk is yellow and the rays white or pink, narrow, and numerous. There is no pappus.

According to Chaucer, the name "daisy" was a contraction of "day's eye," an allusion to its habit of closing at night and not reopening until dawn.

Aster spp., Wild Aster. The Asters make up a large genus of attractive, many-flowered plants. Some difficulty in identification can be expected, but this arises chiefly from the sheer number of species that must be considered. As a group, they may be recognized by a combination of features: fairly narrow ray flowers in white or shades of pink, lavender, violet, or blue (but not yellow), a relatively small disk with yellow or reddish (seldom white) corollas, and overlapping involucral bracts in several series. Most bloom in summer and fall.

The task of identifying individual species can be eased somewhat by marking off groups with shared characteristics, and in this case it is the shapes of the leaves rather than the flowers that afford the best means of segregating them.

Lower leaves both cordate and petioled: *A. cordifolius,*
    A. undulatus, A. macrophyllus, A. schreberi, A. divaricatus
No leaves both cordate and petioled; cauline leaves clasping:
    A. novae-angliae, A. puniceus, A. lucidulus, A. patens,
    A. phlogifolius, A. oblongifolius, A. prenanthoides, A. laevis
No leaves either cordate or clasping; rays colored:
    A. linariifolius, A. concolor, A. curtisii, A. surculosus,
    A. radula, A. prealtus, A. dumosus, A. tataricus
No leaves either cordate or clasping; rays white:
    A. acuminatus, A. umbellatus, A. infirmus, A. pilosus,
    A. lanceolatus, A. lateriflorus, A. paternus, A. solidagineus

Lower leaves both cordate and petioled

Aster cordifolius, Heart-leaved Aster. This and the next two are sometimes difficult to separate because of confusing interchanges of characters. They are large plants with numerous flowers in a paniculate inflorescence; their basal and lower cauline leaves are toothed, petioled, and have cordate bases. The flower heads are $1/2$ to $3/4$ inch wide with 10 to 20 blue or lavender rays. All three grow in rich woods and open areas. In *A. cordifolius* the leaf blades are broadly ovate, and the petioles are slender. In **var. *sagittifolius*** (*A. sagittifolius*), Arrow-leaved Aster (plate 581), the lower leaves have more elongate blades and broadly winged petioles; those near the inflorescence are more lanceolate and nearly sessile. Var. **laevigatus** (*A. lowrieanus*), Lowrie's Aster, may be distinguished by its very smooth, glaucous leaves.

A. undulatus, Wavy-leaved Aster, has a panicle of blue flowers similar to the preceding species. Its leaves are rough, essentially entire, and have wavy margins. The basal ones have cordate bases and are petioled; the upper ones are smaller, and either are sessile with clasping cordate bases or are on auriculate-clasping petioles. The involucral bracts are downy, without glands. Dry, open woods.

A. macrophyllus, Large-leaved Aster, has a creeping rhizome that produces large colonies of sterile plants. Its basal leaves are thick and long-stalked, with a toothed cordate blade that may exceed 8 inches in length; the upper cauline leaves are sessile but not clasping. The flat-topped corymbiform inflorescence is minutely glandular as are the involucres. The rays are pale violet. Dry woods and clearings.

A. schreberi, Schreber's Aster. This and the next are our only two white-flowered species with leaves that are both cordate and petioled. Schreber's Aster resembles *A. macrophyllus*, but it is eglandular and its basal leaves are thinner and have a much broader sinus. The heads ordinarily have from 6 to 14 rays. Woodlands.

A. divaricatus, White Wood Aster (plate 582), is a more delicate plant with a zigzag stem and well-stalked, coarsely sharp-toothed, cordate-ovate leaves. The inflorescence is corymbiform and eglandular. This common species has ³/₄-inch heads with only 5 to 10 white rays, but **var. chlorolepis,** which is found at high elevations in the southern mountains, has larger heads with 12 or more rays. Woodlands.

No leaves both cordate and petioled;
cauline leaves clasping

A. novae-angliae, New England Aster (plate 583). The species in this group have cauline leaves with rounded or auriculate bases that clasp the stem. New England Aster is widespread and frequently cultivated. There are well over 50 ray flowers in each 1- to 2-inch head; their color, whether the typical dark violet or the less common deep rose, is intense. The disk is deep yellow or orange, and the involucres are densely glandular. Moist, open areas.

The name Michaelmas Daisy (chosen because of the timing of the Feast of St. Michael on September 29) is applied to several species of showy native Asters that have been enthusiastically selected by gardeners. Foremost in popularity among them is New England Aster, with the coastal A. novi-belgii, *or New York Aster, a close rival.*

A. puniceus, Purple-stemmed Aster (plate 584), is a tall, handsome plant thriving in moist habitats. It also has numerous rays, but there may be fewer than 50 and they are lighter blue or lavender; the disk flowers are pale yellow or red, and the involucres eglandular. The stems are purplish, with spreading hairs, and the lanceolate leaves are auriculate-clasping and usually toothed. **A. lucidulus** is a similar but mostly smooth plant with pale or occasionally white flowers.

The leaves of **A. patens,** Late Purple Aster (plate 585), are unusual in that they have rounded basal lobes that almost encircle the stem; they are thick and have entire margins. The 1-inch heads have about 20 blue or violet rays, yellow or reddish disk flowers, and glandular involucres. In the similar **A. phlogifolius** (A. *patens* var. *phlogifolius)* the leaves are thinner in texture and often slightly narrowed above the clasping base. Dry, open woods.

A. oblongifolius (*Virgulus oblongifolius)* is a sprawling colonial plant with only slightly auriculate-clasping cauline leaves. Its involucral bracts are densely glandular and have spreading tips. The heads have up to 40 blue or violet rays about $1/2$ inch long. Dry, rocky places.

A. prenanthoides, Crooked-stemmed Aster, has a zigzag stem. Its scabrous leaves have long-pointed, ovate-lanceolate, sharply toothed blades, but taper below the middle to a long, winged petiole-like base that terminates in clasping auricular lobes. The 1-inch heads have 20 to 35 blue or lavender rays. Moist woods.

A. laevis, Smooth Aster, is a glaucous plant with thick, firm foliage. The lower leaves have lanceolate (never cordate) blades, sometimes with a few teeth, tapering to winged petioles; the upper ones are smaller, entire, sessile, and widest at the clasping base. There are 15 to 20 blue or violet rays in each ³/₄-inch head. In **var. concinnus** the leaves are narrower and only slightly clasping, and are not glaucous. Dry, open places.

No leaves either cordate or clasping; rays colored

A. linariifolius, Stiff Aster (plate 586), has several clumped, rigid stems, crowded with stiff, ascending leaves less than 1 ¹/₂ inch long; they are all alike—linear, entire, scabrous, and sessile. The 1-inch flower heads are in a corymbiform inflorescence and have 10 to 20 violet rays. Dry, open habitats.

A. concolor, Eastern Silvery Aster, has its flowers in an elongate raceme. The sessile leaves are under 2 inches long, lanceolate, entire, and silky-hairy on both sides. The blue-violet flower heads are about ³/₄ inch across. Dry, sandy places.

A. curtisii, Curtis' Aster (plate 587) is a southern woodland species with smooth, toothed, elliptic to linear cauline leaves, the basal ones often absent at flowering time. The heads are 1 inch wide, with bright violet rays and yellow to red disks. It can be readily recognized by the bracts of the involucre, which are green, leafy, and strongly recurved.

In **A. surculosus** the principal leaves are scabrous and entire, with elliptic blades tapering to a long petiole, becoming smaller and sessile above; basal leaves are present. There are several flower heads in a corymbiform cluster, each with 15 to 30 violet rays; the involucral bracts are spreading but not recurved. Dry soil.

A. radula, Rough-leaved Aster, has sessile, lanceolate cauline leaves with toothed margins, scabrous above and conspicuously veined. Several 1-inch heads with 15 to 30 violet rays are borne in a flat corymbiform inflorescence. Wet places.

A. prealtus, Tall Aster, has many small flowers in a crowded panicle of leafy branches. The numerous cauline leaves are lanceolate and long-tapered at both ends, sessile, with a conspicuous network of veins on the underside. The heads are small, with 20 or more pale blue-violet rays. Moist, low ground.

A. *dumosus,* Bushy Aster, is a much-branched plant with abundant flowers on long stalks in a spreading inflorescence. The normal leaves are linear, but at flowering time the foliage consists principally of a great many short bractlike leaves on the branches. The heads are $^1/_2$ to $^3/_4$ inch wide and have about 20 pale blue rays. Sandy soil.

A. *tataricus,* Tartarian Aster, an Asiatic introduction escaped from cultivation, cannot be mistaken for any of our native species. It is a coarse, hairy plant with toothed, elliptic to oblong leaves; the lower ones may be 1 inch or more long and one-third as wide. The flower heads are 1 $^1/_2$ inches across, with 15 to 20 lavender rays, in a flat-topped cluster.

No leaves either cordate or clasping; rays white

A. *acuminatus,* Whorled Wood Aster (plate 588). The largest leaves of A. *acuminatus* are closely crowded on the upper part of the stem, forming what may appear to be a whorl below the inflorescence. They are elliptic, coarsely toothed, long-pointed, and taper at the base to an obscure petiole. The flower heads are in an open corymb, with about 15 white rays and a yellow or red disk. Woodlands.

A. *umbellatus,* Flat-topped White Aster (plate 589). The inflorescence of this species is not an umbel, as the name would imply, but a flattish corymbiform cluster. There are about a dozen white rays in each $^3/_4$-inch head; the disk is dull yellow. Its leaves are narrowly elliptic and entire, on short petioles. Moist places.

A. *infirmus,* Cornel-leaved Aster. Compared to the preceding species, this is a smaller plant and has a more open inflorescence, and its heads are fewer and on longer, leafless stalks. There are 5 to 10 rays and a yellow disk. Woodlands.

A. *pilosus,* White Heath Aster (plate 590), is tall, many-branched, and densely pubescent, with linear to narrowly elliptic leaves which may be entire or slightly toothed. The inflorescence is a diffuse panicle invested with a great many small, sharp-pointed bractlike leaves. The heads are $^1/_2$ to $^3/_4$ inch wide, with 20 to 30 rays and yellow or red disks. Fields and waste areas.

A. lanceolatus (*A. simplex*), Panicled Aster, is smooth and has its flowers in an elongate, leafy inflorescence. The heads are $^3/_4$ inch wide, with 20 to 40 rays. Its leaves are narrowly lanceolate, toothed, or entire. Moist, low places.

A. lateriflorus, Calico Aster (plate 591), has unusually small flower heads—between $^1/_4$ and $^1/_2$ inch across—and they frequently are secund on the widely spreading leafy branches of the panicle. There are less than 15 white rays, which occasionally have a lavender tinge, and the disks are purplish red. Woods and open areas.

A. paternus (*Sericocarpus asteroides*), Toothed White-topped Aster (plate 592). The so-called White-topped Asters have $^1/_2$-inch heads with only about 5 rays and creamy-white disks, borne in flat-topped clusters. In this species the leaves are elliptic, the lower ones tapering to a petiole and bearing a few teeth above the middle, the upper ones smaller and sessile. The involucres are cylindrical with squarrose-tipped bracts, and the pappus is brownish. Dry woods.

A. solidagineus (*Sericocarpus linifolius*), Narrow-leaved White-topped Aster, has sessile, entire, linear to narrowly elliptic leaves, not differing very much in size. It has a white pappus. Dry woods and roadsides. Summer.

Erigeron spp. The few species of *Erigeron* occurring in our area can be separated fairly well from the *Asters* on several bases: They begin to flower in the spring (although some continue until fall); their involucral bracts are linear and are arranged in a single series; and usually there are 50 or more very narrow rays in each head.

Erigeron pulchellus, Robin's Plantain (plate 593), is one of the earliest composites to bloom. It is a stoloniferous, hairy plant with oblanceolate, scalloped basal leaves, and a few small alternate leaves with clasping bases above. The flower heads are solitary to several, about 1 inch across, with 50 to 100 lavender or whitish rays. Open woods. Spring.

E. ***philadelphicus,*** Common Fleabane (plate 594), also has clasping upper leaves but more numerous and smaller heads (⁵/₈ inch wide), each with 100 or more very narrow pink to white rays. Open places. Spring–early summer.

E. ***annuus,*** Daisy Fleabane, has long spreading hairs and coarsely toothed cauline leaves more than ³/₈ inch wide, sessile but not clasping. The heads are ¹/₂ to ³/₄ inch wide and have 50 or many more white rays. Fields and other open areas. Spring to fall.

In *E.* ***strigosus,*** Daisy Fleabane (plate 595), the stems have short appressed hairs, and the sessile leaves are narrower and mostly untoothed. Fields, roadsides, and waste places. Spring to fall.

Galinsoga quadriradiata *(G. ciliata)*, Peruvian Daisy (plate 596), is a common weed found blooming in gardens and waste places from late spring until fall. It is a small, hairy plant with opposite, short-stalked, ovate, toothed leaves. The receptacle is conical, less than ¹/₄ inch wide, the disk flowers yellow and fertile. There usually are 5 white rays, ¹/₈ inch wide and long, each with 3 teeth at the apex, and with distinct pappus scales.

G. ***parviflora*** may be separated on the basis of its ray flowers, which are even smaller and without pappus. Also, the pubescence tends to be appressed rather than spreading.

During World War II this unassuming weed enabled the British to indulge in a bit of wordplay. Amid the rubble left by the bombing of London, the first plant to emerge was the brave little Galinsoga, *and they couldn't resist dubbing it "Gallant Soldier."*

Parthenium integrifolium, Wild Quinine (plate 597), is a sturdy plant with thick, rough-surfaced, sparsely toothed alternate leaves, sessile but not clasping. Its flowers are small, in dense flat-topped clusters. The disk is ¹/₄ inch wide, its flowers white and sterile. There are 5 tiny white rays evenly spaced around the perimeter. Open woods, fields, and roadsides. Summer.

Achillea millefolium ssp. lanulosa, Yarrow (plate 598). The alternate, fernlike, pinnately dissected leaves of Yarrow are strongly aromatic. The $1/4$-inch heads of small flowers are in a compact, more or less flat-topped inflorescence, all of them fertile and usually dull white but sometimes pink. There are 5 very short, 3-toothed rays per head. Fields, roadsides, and waste places. Summer.

It is said that Achilles used crushed Yarrow leaves to stanch the flow of blood from his wounds, hence the generic name. Whether or not early American colonists knew of this, they did recall the effectiveness of a tea prepared from it in Europe when used as a febrifuge, and were quick to adopt this indigenous subspecies as a substitute. They also hung Yarrow in their tool sheds to protect against harm, which may have been less a superstition than a reminder to exercise caution, given the proximity of sharp implements.

Conyza canadensis (*Erigeron canadensis*), Horseweed (plate 599), is a coarse, tall weed with numerous linear cauline leaves (the basal ones are absent at flowering time). The inflorescence is a large open panicle bearing a great many small heads. The involucre is $1/8$ inch high, and the whitish rays are narrow, short, and erect. Waste ground. Summer–fall.

Eclipta alba, Yerba de Tajo (plate 600), belongs to a tropical genus and is occasionally found here as a weed in moist waste places, blooming in summer and fall. Its leaves are narrow, long-tapered at both ends, opposite, and held horizontally. The flowers, which are in small axillary and terminal clusters, have a flat receptacle and a 1/4-inch-wide disk of white, 4-toothed, fertile flowers; the rays are also white, narrow, and short.

Plate 1.
*Typha
latifolia*,
Common
Cat-tail

Plate 2.
*Sparganium
americanum*,
Bur-reed

Plate 3.
*Alisma
subcordatum*,
Water
Plantain

Plate 4.
*Sagittaria
latifolia*,
Arrowhead

Plate 5.
*Sagittaria
fasciculata*,
Bunched
Arrowhead

Plate 6.
*Cymophyllus
fraseri*,
Fraser's Sedge

Plate 7.
*Arisaema
triphyllum*,
Jack-in-the-
Pulpit

Plate 8.
*Peltandra
virginica*,
Arrow Arum

Plate 9.
*Symplocarpus
foetidus*,
Skunk
Cabbage

Plate 10.
*Acorus
calamus*,
Sweetflag

Plate 11.
Orontium aquaticum,
Golden
Club

Plate 12.
Xyris torta,
Yellow-eyed
Grass

Plate 13.
Commelina communis,
Asiatic
Dayflower

Plate 14.
Tradescantia ohiensis,
Spiderwort

Plate 15.
Murdannia keisak,
Murdannia

Plate 16.
Pontederia cordata,
Pickerel
Weed

Plate 17.
Yucca filamentosa,
Yucca

Plate 18.
Aletris farinosa,
Colic-root

Plate 19.
Convallaria montana,
Wild Lily-of-
the-Valley

Plate 20.
*Streptopus
roseus*,
Rose
Twisted
Stalk

Plate 21.
*Polygonatum
pubescens*,
Solomon's
Seal

Plate 22.
*Disporum
lanuginosum*,
Yellow
Mandarin

Plate 23.
*Disporum
maculatum*,
Nodding
Mandarin

Plate 24. *Uvularia perfoliata*,
Perfoliate Bellwort

Plate 25.
*Uvularia
grandiflora*,
Large-
flowered
Bellwort

Plate 26.
Erythronium umbilicatum,
Trout Lily

Plate 27.
Clintonia borealis,
Bluebead Lily

Plate 28.
Clintonia umbellulata,
Speckled Wood Lily

Plate 29.
Lilium philadel-phicum,
Wood Lily

Plate 30.
Lilium michauxii,
Carolina Lily

Plate 31. *Lilium grayi*,
Gray's Lily

Plate 32.
*Hemerocallis
fulva*,
Orange
Daylily

Plate 33.
*Medeola
virginiana*,
Indian
Cucumber
Root

Plate 34.
*Maianthemum
canadense*,
Canada
Mayflower

Plate 35.
*Allium
canadense*,
Wild Onion

Plate 36.
*Allium
cernuum*,
Nodding
Wild Onion

Plate 37.
*Nothoscordum
bivalve*, False
Garlic

Plate 38.
Ornithogalum umbellatum,
Star-of-Bethlehem

Plate 39.
Camassia scilloides,
Wild Hyacinth

Plate 40.
Helonias bullata,
Swamp Pink

Plate 41.
Smilacina racemosa,
False Solomon's Seal

Plate 42.
Smilacina stellata,
Starry Solomon's Seal

Plate 43.
Xerophyllum asphodeloides, Turkey Beard

Plate 44.
Amianthium muscae-toxicum, Fly Poison

Plate 45.
Chamaelirium luteum, Fairy Wand

Plate 46.
Stenanthium gramineum, Featherbells

Plate 47.
Melanthium latifolium, Bunchflower

Plate 48.
Veratrum viride, False Hellebore

Plate 49.
Zigadenus glaucus,
White
Camass

Plate 50.
Tofieldia glutinosa,
False
Asphodel

Plate 51.
Trillium grandiflorum,
Large-
flowered
Trillium

Plate 52.
Trillium undulatum,
Painted
Trillium

Plate 53.
Trillium erectum,
Wake Robin

Plate 54.
Trillium erectum,
Wake Robin

Plate 55.
Trillium vaseyi,
Vasey's
Trillium

Plate 56.
Trillium cernuum,
Nodding
Trillium

Plate 57.
*Trillium
catesbaei*,
Catesby's
Trillium

Plate 58.
*Trillium
luteum*,
Yellow
Toadshade

Plate 59.
*Trillium
cuneatum*,
Little Sweet
Betsy

Plate 60. *Smilax herbacea*, Carrion Flower

Plate 61. *Hypoxis hirsuta,* Yellow Star Grass

Plate 62. *Zephyranthes atamasco,* Atamasco Lily

Plate 63.
Hymenocallis caroliniana,
Spider Lily

Plate 64. *Dioscorea villosa*,
Wild Yam

Plate 65. *Iris cristata*,
Crested
Dwarf Iris

Plate 66. *Iris verna* var. *smalliana*, Dwarf Iris

Plate 67. *Iris pseudacorus*, Yellow Flag

Plate 68. *Sisyrinchium angustifolium*, Blue-eyed Grass

Plate 69.
Belamcanda chinensis,
Blackberry Lily

Plate 70. *Cypripedium acaule*,
Pink Lady's Slipper

Plate 71.
Cypripedium calceolus var.
pubescens,
Yellow Lady's Slipper

Plate 72.
*Cypripedium
reginae*,
Showy
Lady's
Slipper

Plate 73.
*Arethusa
bulbosa*, Bog
Rose

Plate 74.
*Pogonia
ophio-
glossoides*,
Rose
Pogonia

Plate 75.
*Cleistes
divaricata*,
Spreading
Pogonia

Plate 76.
*Triphora
trianthophora*,
Three Birds
Orchid

Plate 77.
*Isotria
verticillata*,
Whorled
Pogonia

Plate 78.
*Isotria
medeoloides*,
Small
Whorled
Pogonia

Plate 79.
*Aplectrum
hyemale*,
Puttyroot

Plate 80.
*Tipularia
discolor*,
Cranefly
Orchid

Plate 81.
*Corallorhiza
wisteriana*,
Spring
Coralroot

Plate 82.
*Calopogon
tuberosus*,
Grass Pink

Plate 83.
*Galearis
spectabilis*,
Showy
Orchis

Plate 84.
*Listera
smallii*,
Appala-
chian
Twayblade

Plate 85.
*Liparis
lilifolia*,
Lily-leaved
Twayblade

Plate 86.
*Malaxis
unifolia*,
Green
Adder's
Mouth

Plate 87.
*Platanthera
grandiflora*,
Large Purple
Fringed
Orchid

Plate 88.
*Platanthera
ciliaris*,
Yellow
Fringed
Orchid

Plate 89.
*Platanthera
blephariglottis*,
Large White
Fringed
Orchid

Plate 90.
*Platanthera
lacera*,
Ragged
Fringed
Orchid

Plate 91.
*Platanthera
orbiculata*,
Large Round-
leaved
Orchid

Plate 92.
*Platanthera
clavellata*,
Small Green
Wood
Orchid

Plate 93.
*Spiranthes
cernua*,
Nodding
Ladies'
Tresses

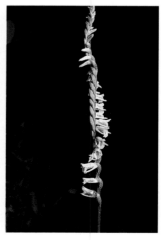

Plate 94.
*Spiranthes
lacera* var.
gracilis
Slender
Ladies'
Tresses

Plate 95.
Goodyera pubescens,
Downy
Rattlesnake
Plantain

Plate 96.
Goodyera repens var.
ophioides,
Lesser
Rattlesnake
Plantain

Plate 97.
Saururus cernuus,
Lizard's Tail

Plate 98. *Comandra umbellata*,
Bastard Toadflax

Plate 99.
Aristolochia macrophylla,
Dutchman's Pipe

Plate 100.
Asarum canadense,
Wild Ginger

Plate 101.
Hexastylis arifolia, Little Brown Jugs

Plate 102.
Hexastylis heterophylla,
Heartleaf

Plate 103.
Polygonum cuspidatum,
Japanese Knotweed

Plate 104.
Polygonum sagittatum,
Arrow-leaved Tearthumb

Plate 105.
Polygonum cilinode,
Fringed Bindweed

Plate 106.
*Polygonum
pensylvanicum*,
Pink
Smartweed

Plate 107.
*Polygonum
cespitosum*
var.
longisetum,
Long-bristled
Smartweed

Plate 108.
*Mirabilis
nyctaginea*,
Wild Four-
o'clock

Plate 109.
*Phytolacca
americana*,
Pokeweed

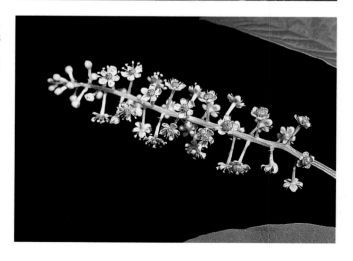

Plate 110.
*Claytonia
caroliniana*,
Spring
Beauty

Plate 111.
*Claytonia
virginica*,
Spring
Beauty

Plate 112. *Talinum teretifolium*,
Fameflower

Plate 113.
Silene latifolia
ssp. *alba*,
White
Campion

Plate 114.
*Silene
vulgaris*,
Bladder
Campion

Plate 115.
Silene stellata,
Starry
Campion

Plate 116. *Silene ovata*, Fringed
Campion

Plate 117.
*Silene
caroliniana*
ssp.
pensylvanica,
Wild Pink

Plate 118.
*Silene
virginica*, Fire
Pink

Plate 119.
*Lychnis flos-
cuculi*,
Ragged
Robin

Plate 120.
*Agrostemma
githago*, Corn
Cockle

Plate 121.
Saponaria
officinalis,
Bouncing Bet

Plate 122.
Dianthus
armeria,
Deptford
Pink

Plate 123.
Cerastium
arvense, Field
Chickweed

Plate 124.
Cerastium fontanum ssp. *triviale*, Mouse-ear Chickweed

Plate 125.
Stellaria pubera, Giant Chickweed

Plate 126.
Stellaria graminea, Lesser Stitchwort

Plate 127.
Minuartia groenlandica,
Mountain Sandwort

Plate 128.
Nymphaea odorata,
Fragrant Water Lily

Plate 129. *Nuphar luteum* ssp.
macrophyllum, Spatterdock

Plate 130.
*Aconitum
uncinatum*,
Monkshood

Plate 131.
*Delphinium
tricorne*,
Dwarf
Larkspur

Plate 132.
*Xanthorhiza
simplicissima*,
Yellowroot

Plate 133. *Aquilegia canadensis*,
Columbine

Plate 134.
Hepatica americana,
Round-lobed
Hepatica

Plate 135.
Hepatica acutiloba,
Sharp-lobed
Hepatica

Plate 136.
Clematis virginiana,
Virgin's
Bower

Plate 137.
*Clematis
viorna*,
Leatherflower

Plate 138.
*Anemone
virginiana*,
Thimble-
weed

Plate 139.
*Anemone
canadensis*,
Canada
Anemone

Plate 140.
*Anemone
quinquefolia*,
Wood
Anemone

Plate 141.
*Coptis
groenlandica*,
Goldthread

Plate 142.
*Caltha
palustris*,
Marsh
Marigold

Plate 143. *Ranunculus
abortivus*, Kidney-leaved
Crowfoot

Plate 144.
*Ranunculus
recurvatus*,
Hooked
Buttercup

Plate 145.
*Ranunculus
acris*, Tall
Buttercup

Plate 146.
*Ranunculus
hispidus*,
Hispid
Buttercup

Plate 147.
*Ranunculus
septen-
trionalis*,
Swamp
Buttercup

Plate 148.
Hydrastis canadensis,
Golden Seal

Plate 149.
Actaea pachypoda,
White Baneberry

Plate 150.
Actaea pachypoda,
White Baneberry

Plate 151.
Cimicifuga racemosa,
Black Cohosh

Plate 152. *Trautvetteria carolinensis*, Tassel Rue

Plate 153. *Thalictrum clavatum*, Mountain Meadow Rue

Plate 154. *Thalictrum dioicum*, Early Meadow Rue

Plate 155. *Thalictrum pubescens*, Tall Meadow Rue

Plate 156. *Thalictrum thalictroides*, Rue Anemone

Plate 157. *Podophyllum peltatum*, May Apple

Plate 158. *Diphylleia cymosa*, Umbrella Leaf

Plate 159.
Jeffersonia diphylla,
Twinleaf

Plate 160.
Caulophyllum thalictroides,
Blue Cohosh

Plate 161.
Papaver dubium,
Poppy

Plate 162.
Chelidonium majus,
Celandine

Plate 163.
Stylophorum diphyllum,
Celandine
Poppy

Plate 164.
Sanguinaria canadensis,
Bloodroot

Plate 165.
*Argemone
mexicana*,
Prickly
Poppy

Plate 166.
*Dicentra
cucullaria*,
Dutchman's
Breeches

Plate 167.
*Dicentra
canadensis*,
Squirrel
Corn

Plate 168.
*Dicentra
eximia*, Wild
Bleeding
Heart

Plate 169.
Adlumia fungosa,
Climbing
Fumitory

Plate 170. *Corydalis flavula*,
Yellow Fumitory

Plate 171.
*Corydalis
sempervirens*,
Pale
Corydalis

Plate 172. *Barbarea vulgaris*,
Winter Cress

Plate 173.
*Raphanus
raphanistrum*,
Wild Radish

Plate 174. *Isatis tinctoria*, Dyer's Woad

Plate 175. *Erysimum cheiranthoides*, Wormseed Mustard

Plate 176. *Capsella bursa-pastoris*, Shepherd's Purse

Plate 177. *Arabis laevigata*, Smooth Rockcress

Plate 178.
Cardamine
hirsuta, Hairy
Bitter Cress

Plate 179. *Cardamine*
pensylvanica, Pennsylvania
Bitter Cress

Plate 180.
Cardamine
clematitis,
Bitter Cress

Plate 181.
*Cardamine
concatenata*,
Cut-leaved
Toothwort

Plate 182.
*Nasturtium
officinale*,
Water Cress

Plate 183.
*Alliaria
petiolata*,
Garlic
Mustard

Plate 184.
*Hesperis
matronalis*,
Dame's
Rocket

Plate 185. *Lunaria annua,*
Honesty

Plate 186.
*Sarracenia
purpurea,*
Pitcher Plant

Plate 187.
*Sarracenia
purpurea,*
Pitcher Plant

Plate 188.
Sarracenia flava,
Yellow Pitcher
Plant

Plate 189.
*Drosera
rotundifolia*,
Round-
leaved
Sundew

Plate 190.
*Drosera
intermedia*,
Spatulate-
leaved
Sundew

Plate 191.
*Sedum
ternatum*,
Wild
Stonecrop

Plate 192.
*Sedum
smallii*, Elf
Orpine

Plate 193. *Sedum telephioides*,
Wild Live-forever

Plate 194.
*Saxifraga
michauxii*,
Michaux's
Saxifrage

Plate 195.
*Saxifraga
michauxii*,
Michaux's
Saxifrage

Plate 196.
Saxifraga micran-thidifolia,
Mountain Lettuce

Plate 197.
Saxifraga virginiensis,
Early Saxifrage

Plate 198.
Saxifraga pensylvanica,
Swamp Saxifrage

Plate 199.
Heuchera villosa,
Alumroot

Plate 200.
Heuchera villosa,
Alumroot

Plate 201.
*Mitella
diphylla,*
Miterwort

Plate 202.
*Mitella
diphylla,*
Miterwort

Plate 203.
*Tiarella
cordifolia,*
Foamflower

Plate 204.
*Astilbe
biternata,*
False
Goatsbeard

Plate 205.
*Parnassia
asarifolia,*
Grass-of-
Parnassus

Plate 206.
Penthorum sedoides,
Ditch
Stonecrop

Plate 207.
Chryso-splenium americanum,
Golden
Saxifrage

Plate 208. *Geum canadense*,
White Avens

Plate 209.
*Geum
radiatum*,
Spreading
Avens

Plate 210. *Agrimonia parviflora*,
Small-flowered Agrimony

Plate 211.
*Potentilla
simplex*,
Common
Cinquefoil

Plate 212. *Potentilla canadensis*, Dwarf Cinquefoil

Plate 213. *Potentilla norvegica*, Rough Cinquefoil

Plate 214. *Potentilla recta*, Rough-fruited Cinquefoil

Plate 215.
*Sibbaldiopsis
tridentata*,
Wine-leaved
Cinquefoil

Plate 216.
*Fragaria
virginiana*,
Wild
Strawberry

Plate 217.
*Duchesnea
indica*, Indian
Strawberry

Plate 218. *Waldsteinia fragarioides*, Barren Strawberry

Plate 219. *Dalibarda repens*, Dewdrop

Plate 220. *Rubus odoratus*, Purple-flowering Raspberry

Plate 221.
Porteranthus trifoliatus,
Bowman's Root

Plate 222. *Filipendula rubra*,
Queen-of-the-Prairie

Plate 223.
Aruncus dioicus,
Goatsbeard

Plate 224. *Sanguisorba canadensis*, American Burnet

Plate 225. *Schrankia microphylla*, Sensitive Brier

Plate 226. *Chamaecrista fasciculata*, Partridge Pea

Plate 227.
*Chamaecrista
nictitans*,
Wild
Sensitive
Plant

Plate 228. *Lathyrus venosus*,
Veiny Pea

Plate 229.
*Vicia
angustifolia*,
Narrow-
leaved Vetch

Plate 230.
*Vicia
dasycarpa*,
Smooth
Vetch

Plate 231.
*Vicia
caroliniana*,
Wood Vetch

Plate 232.
*Clitoria
mariana*,
Butterfly Pea

Plate 233.
*Centrosema
virginianum*,
Spurred
Butterfly Pea

Plate 234.
*Pueraria
lobata*, Kudzu

Plate 235.
*Pueraria
lobata*, Kudzu

Plate 236.
*Phaseolus
polystachios*,
Wild Bean

Plate 237.
*Strophostyles
helvula*,
Trailing Wild
Bean

Plate 238. *Amphicarpaea
bracteata*, Hog Peanut

Plate 239.
*Galactia
regularis*,
Milk Pea

Plate 240.
*Baptisia
tinctoria*,
Wild Indigo

Plate 241.
Baptisia alba,
White Wild
Indigo

Plate 242.
*Thermopsis
villosa*, Hairy
Bush Pea

Plate 243.
*Stylosanthes
biflora*, Pencil
Flower

Plate 244. *Psoralea psoralioides*
var. *eglandulosa*, Sampson's
Snakeroot

Plate 245.
*Desmodium
rotundifolium*,
Prostrate
Tick Trefoil

Plate 246. *Desmodium paniculatum*, Panicled Tick Trefoil

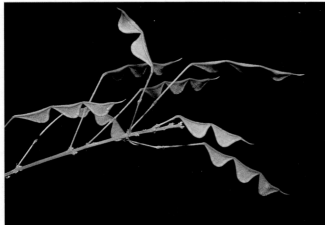

Plate 247. *Desmodium nudiflorum*, Naked-flowered Tick Trefoil

Plate 248. *Lespedeza repens*, Creeping Bush Clover

Plate 249.
Lespedeza intermedia,
Wandlike
Bush Clover

Plate 250.
Lespedeza capitata,
Round-headed Bush
Clover

Plate 251.
Lespedeza cuneata,
Sericea

Plate 252.
Trifolium aureum, Hop
Clover

Plate 253.
Trifolium repens,
White
Clover

Plate 254.
Trifolium hybridum,
Alsike
Clover

Plate 255.
Trifolium virginicum, Kates Mountain Clover

Plate 256.
Trifolium pratense, Red Clover

Plate 257.
Trifolium incarnatum, Crimson Clover

Plate 258.
Trifolium arvense, Rabbit Foot Clover

Plate 259. *Medicago lupulina*, Black Medick

Plate 260. *Melilotus officinalis*, Yellow Sweet Clover

Plate 261. *Apios americana*, Groundnut

Plate 262. *Lotus corniculatus*, Bird's-foot Trefoil

Plate 263. *Coronilla varia*, Crown Vetch

Plate 264.
*Tephrosia
virginiana*,
Goat's Rue

Plate 265.
*Astragalus
canadensis*,
Milk Vetch

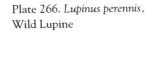

Plate 266. *Lupinus perennis*,
Wild Lupine

Plate 267.
*Oxalis
montana*,
Common
Wood Sorrel

Plate 268.
*Oxalis
violacea*,
Violet Wood
Sorrel

Plate 269.
*Oxalis
grandis*, Large
Wood Sorrel

Plate 270.
*Geranium
maculatum*,
Wild
Geranium

Plate 271.
*Geranium
sibiricum*,
Siberian
Cranesbill

Plate 272. *Geranium carolinianum*, Carolina Cranesbill

Plate 273. *Geranium robertianum*, Herb Robert

Plate 274. *Erodium cicutarium*, Storksbill

Plate 275. *Linum striatum*,
Wild Flax

Plate 276.
*Polygala
paucifolia*,
Fringed
Polygala

Plate 277.
*Polygala
senega*,
Seneca
Snakeroot

Plate 278.
*Polygala
curtissii*,
Curtiss'
Milkwort

Plate 279.
Euphorbia corollata,
Flowering Spurge

Plate 280.
Euphorbia purpurea,
Glade Spurge

Plate 281.
Euphorbia cyparissias,
Cypress Spurge

Plate 282.
Pachysandra procumbens,
Allegheny Spurge

Plate 283.
Euonymus americanus,
Strawberry
Bush

Plate 284.
Euonymus obovatus,
Trailing
Strawberry
Bush

Plate 285.
Impatiens capensis,
Spotted
Jewel Weed

Plate 286.
*Impatiens
pallida*, Pale
Jewel Weed

Plate 287.
*Hibiscus
moscheutos*
ssp. *palustris*,
Swamp Rose
Mallow

Plate 288.
*Malva
moschata*,
Musk
Mallow

Plate 289.
*Malva
neglecta*,
Common
Mallow

Plate 290.
Sida spinosa,
Prickly
Mallow

Plate 291.
*Hypericum
crux-andreae*,
St. Peter's-
wort

Plate 292.
*Hypericum
hypericoides*,
St. Andrew's
Cross

Plate 293.
*Hypericum
graveolens*,
Mountain St.
John's-wort

Plate 294.
*Hypericum
perforatum*,
Common St.
John's-wort

Plate 295. *Hypericum buckleyi*,
Appalachian St. John's-wort

Plate 296.
*Hypericum
gentianoides*,
Pineweed

Plate 297. *Triadenum virginicum*, Marsh St. John's-wort

Plate 298. *Helianthemum canadense*, Frostweed

Plate 299. *Lechea racemulosa*, Pinweed

Plate 300. *Viola rostrata*, Long-spurred Violet

Plate 301.
Viola rafinesquii,
Wild Pansy

Plate 302.
Viola canadensis,
Canada Violet

Plate 303.
Viola pubescens,
Downy Yellow Violet

Plate 304.
Viola tripartita,
Three-parted Violet

Plate 305.
Viola hastata,
Halberd-leaved Violet

Plate 306.
Viola pedata
var.
lineariloba,
Birdfoot
Violet

Plate 307.
Viola palmata
forma *striata*,
Early Blue
Violet

Plate 308.
Viola
fimbriatula,
Downy
Violet

Plate 309. *Viola sororia,* Common Blue Violet

Plate 310. *Viola cucullata,* Marsh Blue Violet

Plate 311. *Viola blanda,* Sweet White Violet

Plate 312.
*Viola
rotundifolia*,
Round-
leaved Yellow
Violet

Plate 313. *Hybanthus concolor*,
Green Violet

Plate 314.
*Passiflora
incarnata*,
Passion
Flower

Plate 315. *Passiflora lutea*, Yellow Passion Flower

Plate 316. *Opuntia humifusa*, Prickly Pear

Plate 317. *Lythrum salicaria*, Purple Loosestrife

Plate 318.
*Cuphea
viscosissima*,
Clammy
Cuphea

Plate 319.
*Rhexia
virginica*,
Meadow
Beauty

Plate 320.
*Oenothera
speciosa*,
Showy
Evening
Primrose

Plate 321. *Oenothera biennis*,
Common Evening Primrose

Plate 322.
Oenothera
fruticosa,
Sundrops

Plate 323.
Ludwigia
decurrens,
Primrose
Willow

Plate 324.
Ludwigia
alternifolia,
Seedbox

Plate 325. *Epilobium*
angustifolium, Fireweed

Plate 326.
*Epilobium
coloratum*,
Purple-leaved
Willow Herb

Plate 327. *Circaea alpina*,
Smaller Enchanter's
Nightshade

Plate 328.
*Gaura
biennis*,
Biennial
Gaura

Plate 329. *Panax quinquefolius*, Ginseng

Plate 330. *Panax trifolius*, Dwarf Ginseng

Plate 331. *Aralia nudicaulis*, Wild Sarsaparilla

Plate 332. *Aralia racemosa*, Spikenard

Plate 333. *Hydrocotyle americana*, Pennywort

Plate 334.
Eryngium yuccifolium,
Rattlesnake Master

Plate 335.
Sanicula gregaria,
Clustered Snakeroot

Plate 336.
Heracleum maximum,
Cow Parsnip

Plate 337.
Pastinaca sativa, Wild
Parsnip

Plate 338.
Oxypolis rigidior,
Cowbane

Plate 339.
Daucus carota, Wild
Carrot

Plate 340. *Chaerophyllum tainturieri*, Wild Chervil

Plate 341. *Osmorhiza longistylis*, Aniseroot

Plate 342. *Angelica triquinata*, Filmy Angelica

Plate 343. *Ligusticum canadense*, American Lovage

Plate 344. *Cicuta maculata*,
Water Hemlock

Plate 345.
*Taenidia
integerrima*,
Yellow
Pimpernel

Plate 346.
Zizia aptera,
Heart-leaf
Alexanders

Plate 347.
*Thaspium
trifoliatum*,
Meadow
Parsnip

Plate 348. *Thaspium barbinode*,
Hairy-jointed Meadow Parsnip

Plate 349.
*Cornus
canadensis*,
Bunchberry

Plate 350.
*Chimaphila
maculata*,
Spotted
Wintergreen

Plate 351.
*Pyrola
rotundifolia*
var.
americana,
Round-
leaved Pyrola

Plate 352.
*Epigaea
repens*,
Trailing
Arbutus

Plate 353.
*Gaultheria
procumbens*,
Wintergreen

Plate 354.
*Monotropa
uniflora*,
Indian Pipe

Plate 355.
*Monotropa
hypopithys*,
Pinesap

Plate 356.
Galax aphylla,
Galax

Plate 357.
*Shortia
galacifolia*,
Oconee Bells

Plate 358. *Lysimachia
quadrifolia*, Whorled
Loosestrife

Plate 359.
*Lysimachia
nummularia*,
Moneywort

Plate 360. *Lysimachia terrestris*,
Swamp Candles

Plate 361.
*Lysimachia
lanceolata*,
Lance-leaf
Loosestrife

Plate 362.
*Anagallis
arvensis,*
Scarlet
Pimpernel

Plate 363.
*Trientalis
borealis,* Star
Flower

Plate 364. *Dodecatheon meadia,*
Shooting Star

Plate 365.
*Spigelia
marilandica*,
Indian Pink

Plate 366.
*Gentianopsis
crinita*,
Fringed
Gentian

Plate 367.
*Gentianella
quinquefolia*,
Stiff
Gentian

Plate 368.
*Gentiana
clausa*,
Closed
Gentian

Plate 369.
*Gentiana
decora*,
Striped
Gentian

Plate 370. *Obolaria virginica*,
Pennywort

Plate 371.
*Frasera
carolinensis*,
American
Columbo

Plate 372.
*Sabatia
angularis*,
Rose Pink

Plate 373.
*Menyanthes
trifoliata*,
Buckbean

Plate 374.
*Apocynum
androsaemi-
folium*,
Spreading
Dogbane

Plate 375. *Amsonia tabernaemontana*, Blue Star

Plate 376. *Vinca minor,* Periwinkle

Plate 377. *Asclepias tuberosa,* Butterfly Weed

Plate 378.
*Asclepias
quadrifolia*,
Four-leaved
Milkweed

Plate 379.
*Asclepias
incarnata* ssp.
pulchra,
Swamp
Milkweed

Plate 380.
*Asclepias
variegata*,
White
Milkweed

Plate 381.
*Asclepias
exaltata*, Poke
Milkweed

Plate 382.
*Asclepias
exaltata*, Poke
Milkweed

Plate 383.
*Asclepias
verticillata*,
Whorled
Milkweed

Plate 384.
*Matelea
carolinensis*,
Climbing
Milkweed

Plate 385. *Calystegia sepium*, Hedge Bindweed

Plate 386. *Convolvulus arvensis*, Field Bindweed

Plate 387. *Ipomoea purpurea*, Morning Glory

Plate 388.
Ipomoea pandurata,
Wild Potato Vine

Plate 389.
Ipomoea lacunosa,
Small White Morning Glory

Plate 390.
Ipomoea coccinea, Red Morning Glory

Plate 391.
*Cuscuta
rostrata*,
Dodder

Plate 392.
*Phlox
subulata*,
Moss Phlox

Plate 393.
*Phlox
stolonifera*,
Creeping
Phlox

Plate 394.
*Phlox
divaricata*,
Wild Blue
Phlox

Plate 395. *Phlox carolina*,
Carolina Phlox

Plate 396.
*Polemonium
reptans*,
Jacob's
Ladder

Plate 397.
Hydrophyllum virginianum,
Waterleaf

Plate 398.
Hydrophyllum macrophyllum,
Large-leaf
Waterleaf

Plate 399.
Phacelia fimbriata,
Fringed
Phacelia

Plate 400.
Phacelia dubia, Small-flowered
Phacelia

Plate 401.
*Phacelia
bipinnatifida*,
Purple
Phacelia

Plate 402. *Echium vulgare*,
Viper's Bugloss

Plate 403.
*Mertensia
virginica*,
Virginia
Bluebell

Plate 404.
Myosotis
scorpioides,
Forget-me-
not

Plate 405.
Lithospermum
canescens,
Hoary
Puccoon

Plate 406.
Onosmodium
virginianum,
False
Gromwell

Plate 407.
Verbena
hastata, Blue
Vervain

Plate 408.
*Verbena
urticifolia*,
White
Vervain

Plate 409.
*Phyla
lanceolata*,
Fog Fruit

Plate 410.
*Monarda
fistulosa*,
Wild
Bergamot

Plate 411.
*Monarda
didyma*,
Oswego Tea

Plate 412.
Satureja vulgaris,
Wild Basil

Plate 413.
Pycnan-themum incanum,
Hoary Mountain Mint

Plate 414.
Pycnan-themum tenuifolium,
Narrow-leaved Mountain Mint

Plate 415.
Teucrium canadense,
American Germander

Plate 416.
*Agastache
scrophulariae-
folia*, Purple
Giant
Hyssop

Plate 417.
*Physostegia
virginiana*,
Obedient
Plant

Plate 418.
*Mentha
piperita*,
Peppermint

Plate 419. *Prunella vulgaris*,
Self Heal

Plate 420.
*Galeopsis
tetrahit*,
Hemp
Nettle

Plate 421. *Leonurus cardiaca*,
Motherwort

Plate 422.
*Lycopus
virginicus*,
Bugleweed

Plate 423.
Marrubium vulgare,
Horehound

Plate 424.
Lamium amplexicaule,
Henbit

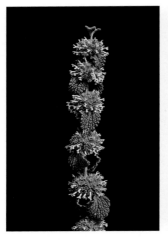

Plate 425.
Glechoma hederacea,
Ground Ivy

Plate 426.
Collinsonia canadensis,
Horse Balm

Plate 427.
Trichostema dichotomum,
Blue Curls

Plate 428.
Stachys latidens,
Hedge Nettle

Plate 429.
Meehania cordata,
Meehania

Plate 430.
Salvia lyrata,
Lyre-leaved Sage

Plate 431. *Scutellaria integrifolia*, Hyssop Skullcap

Plate 432. *Physalis heterophylla*,
Clammy Ground Cherry

Plate 433.
*Datura
stramonium*,
Jimson Weed

Plate 434.
*Solanum
carolinense*,
Horse Nettle

Plate 435.
Solanum rostratum,
Buffalo Bur

Plate 436.
Solanum dulcamara,
Bittersweet
Nightshade

Plate 437.
Verbascum thapsus,
Woolly
Mullein

Plate 438.
Verbascum blattaria,
Moth
Mullein

Plate 439.
Aureolaria virginica,
False
Foxglove

Plate 440.
Agalinis
purpurea,
Purple
Gerardia

Plate 441. *Penstemon digitalis*,
Foxglove Beardtongue

Plate 442.
Penstemon
smallii,
Small's
Penstemon

Plate 443.
*Penstemon
canescens*,
Gray
Beardtongue

Plate 444.
*Chelone
lyonii*,
Turtlehead

Plate 445.
*Chelone
glabra*,
White
Turtlehead

Plate 446.
*Mimulus
ringens*,
Monkey
Flower

Plate 447.
*Pedicularis
canadensis*,
Wood
Betony

Plate 448.
*Scrophularia
marilandica*,
Figwort

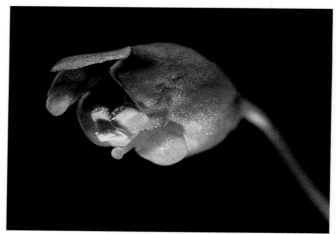

Plate 449. *Veronicastrum
virginicum*, Culver's Root

Plate 450.
Castilleja coccinea,
Indian Paintbrush

Plate 451.
Linaria vulgaris,
Common Toadflax

Plate 452.
Nuttalanthus canadensis,
Old-field Toadflax

Plate 453.
Melampyrum lineare, Cow Wheat

Plate 454.
Veronica officinalis,
Common Speedwell

Plate 455.
*Veronica
persica*,
Bird's-eye
Speedwell

Plate 456.
*Gratiola
neglecta*,
Hedge
Hyssop

Plate 457.
*Lindernia
dubia*, False
Pimpernel

Plate 458.
*Lindernia
monticola*,
False
Pimpernel

Plate 459. *Buchnera americana*,
Bluehearts

Plate 460.
*Bacopa
caroliniana*,
Water
Hyssop

Plate 461.
Mazus
japonica,
Mazus

Plate 462.
Bignonia
capreolata,
Cross Vine

Plate 463.
Campsis
radicans,
Trumpet
Creeper

Plate 464.
Conopholis
americana,
Squaw Root

Plate 465. *Epifagus virginiana*,
Beech Drops

Plate 466.
*Orobanche
uniflora*,
One-
flowered
Cancer Root

Plate 467.
*Justicia
americana*,
Water
Willow

Plate 468.
*Ruellia
caroliniensis*,
Wild
Petunia

Plate 469. *Phryma leptostachya*,
Lopseed

Plate 470.
Galium
verum,
Yellow
Bedstraw

Plate 471.
Galium
asprellum,
Rough
Bedstraw

Plate 472.
Galium
latifolium,
Purple
Bedstraw

Plate 473.
*Houstonia
serpyllifolia*,
Thyme-
leaved
Bluets

Plate 474.
*Houstonia
pusilla*, Small
Bluets

Plate 475.
*Houstonia
purpurea*,
Purple
Bluets

Plate 476. *Diodia teres*,
Buttonweed

Plate 477.
*Diodia
virginiana*,
Buttonweed

Plate 478.
*Mitchella
repens*,
Partridge
Berry

Plate 479.
*Triosteum
aurantiacum*,
Horse
Gentian

Plate 480.
*Valerianella
radiata*, Corn
Salad

Plate 481.
*Dipsacus
sylvestris*,
Teasel

Plate 482.
*Sicyos
angulatus*,
Bur
Cucumber

Plate 483.
*Echinocystis
lobata*,
Balsam
Apple

Plate 484.
*Melothria
pendula*,
Creeping
Cucumber

Plate 485. *Campanula
americana*, Tall Bellflower

Plate 486.
*Campanula
divaricata*,
Southern
Harebell

Plate 487.
*Lobelia
siphilitica*,
Great Blue
Lobelia

Plate 488.
*Lobelia
puberula*,
Downy
Lobelia

Plate 489.
*Lobelia
inflata*,
Indian
Tobacco

Plate 490.
*Lobelia
cardinalis*,
Cardinal
Flower

Plate 491. *Triodanis perfoliata*,
Venus' Looking Glass

Plate 492.
*Prenanthes
roanensis*,
Rattlesnake
Root

Plate 493.
*Cichorium
intybus*,
Chicory

Plate 494. *Lactuca serriola* var. *integrata*, Prickly Lettuce

Plate 495. *Lactuca floridana*, Woodland Lettuce

Plate 496. *Tragopogon porrifolius*, Salsify

Plate 497.
*Hieracium
pilosella*,
Mouse-ear
Hawkweed

Plate 498.
*Hieracium
aurantiacum*,
Devil's
Paintbrush

Plate 499.
*Hieracium
caespitosum*,
King Devil

Plate 500.
*Hieracium
venosum*,
Rattlesnake
Weed

Plate 501.
Taraxacum officinale,
Common Dandelion

Plate 502.
Taraxacum officinale,
Common Dandelion

Plate 503.
Krigia montana,
Mountain Cynthia

Plate 504.
Hypochoeris radicata,
Cat's Ear

Plate 505.
Pyrrhopappus carolinianus,
False Dandelion

Plate 506.
Sonchus asper, Spiny-leaved Sow Thistle

Plate 507.
Crepis capillaris,
Smooth Hawk's Beard

Plate 508.
Lapsana communis,
Nipplewort

Plate 509.
Antennaria solitaria,
Solitary Pussytoes

Plate 510.
Antennaria plantagini-folia,
Plantain-leaved Pussytoes

Plate 511.
Anaphalis margaritacea,
Pearly Everlasting

Plate 512.
Mikania scandens,
Climbing Hempweed

Plate 513. *Ageratina altissima*,
White Snakeroot

Plate 514.
Eupatorium
perfoliatum,
Boneset

Plate 515. *Eupatorium*
fistulosum, Joe-Pye Weed

Plate 516.
Eupatorium coelestinum,
Mistflower

Plate 517.
Erechtites hieracifolia,
Pilewort

Plate 518.
Arnoglossum atriplicifolium,
Pale Indian Plantain

Plate 519.
Rugelia nudicaulis,
Rugel's Indian Plantain

Plate 520.
*Tanacetum
vulgare*,
Tansy

Plate 521.
*Matricaria
matricarioides*,
Pineapple
Weed

Plate 522.
*Cirsium
vulgare*, Bull
Thistle

Plate 523. *Cirsium arvense*,
Canada Thistle

Plate 524. *Cirsium muticum*, Swamp Thistle

Plate 525. *Cirsium altissimum*, Tall Thistle

Plate 526. *Cirsium horridulum*, Yellow Thistle

Plate 527. *Carduus nutans*, Nodding Thistle

Plate 528.
*Arctium
minus*,
Common
Burdock

Plate 529.
*Centaurea
cyanus*,
Bachelor's
Button

Plate 530.
*Centaurea
maculosa*,
Spotted
Knapweed

Plate 531.
*Elephantopus
carolinianus,*
Elephant's
Foot

Plate 532.
*Pluchea
camphorata,*
Camphorweed

Plate 533.
*Gamochaeta
purpurea,*
Purple
Cudweed

Plate 534.
*Liatris
spicata,*
Dense
Blazing Star

Plate 535.
*Liatris
aspera,*
Rough
Blazing Star

Plate 536.
Vernonia noveboracensis,
New York
Ironweed

Plate 537.
Marshallia obovata,
Barbara's
Buttons

Plate 538.
Rudbeckia hirta,
Black-eyed
Susan

Plate 539.
Rudbeckia laciniata,
Green-headed Coneflower

Plate 540.
Helenium autumnale,
Sneezeweed

Plate 541.
Helenium flexuosum,
Purple-headed Sneezeweed

Plate 542.
Helenium
amarum,
Bitterweed

Plate 543. *Helianthus annuus*,
Common Sunflower

Plate 544.
Helianthus
atrorubens,
Hairy Wood
Sunflower

Plate 545.
*Helianthus
microcephalus*,
Small Wood
Sunflower

Plate 546.
*Helianthus
tuberosus*,
Jerusalem
Artichoke

Plate 547.
*Helianthus
resinosus*,
Gray
Sunflower

Plate 548.
*Heliopsis
helianthoides*,
Ox-eye

Plate 549.
*Tetrago-
notheca
helianthoides*,
Pineland
Ginseng

Plate 550.
Coreopsis
tinctoria,
Calliopsis

Plate 551.
Coreopsis
major,
Greater
Tickseed

Plate 552.
Coreopsis
auriculata,
Eared
Coreopsis

Plate 553.
Bidens
polylepis,
Bur Marigold

Plate 554.
Bidens polylepis,
Bur Marigold

Plate 555.
Bidens tripartita,
Beggar Ticks

Plate 556.
Bidens frondosa,
Beggar Ticks

Plate 557.
Pityopsis graminifolia,
Grass-leaved
Golden
Aster

Plate 558. *Chrysopsis mariana*,
Maryland Golden Aster

Plate 559.
*Heterotheca
subaxillaris*,
Camphorweed

Plate 560.
*Senecio
anonymus*,
Squaw Weed

Plate 561. *Senecio tomentosus*,
Woolly Ragwort

Plate 562. *Senecio aureus*,
Golden Ragwort

Plate 563.
*Chrysogonum
virginianum*,
Green-and-
Gold

Plate 564.
*Silphium
compositum*,
Rosinweed

Plate 565.
*Verbesina
occidentalis*,
Crown-beard

Plate 566.
*Verbesina
virginica*,
Tickweed

Plate 567.
*Smallanthus
uvedalia*,
Bearsfoot

Plate 568.
*Tussilago
farfara*,
Coltsfoot

Plate 569.
*Solidago
bicolor*,
Silverrod

Plate 570.
Solidago caesia, Blue-stem Goldenrod

Plate 571.
Solidago glomerata, Skunk Goldenrod

Plate 572.
Solidago roanensis, Mountain Goldenrod

Plate 573.
Solidago nemoralis, Gray Goldenrod

Plate 574. *Solidago arguta*, Sharp-leaved Goldenrod

Plate 575.
*Solidago
odora*, Sweet
Goldenrod

Plate 576.
*Solidago
canadensis*,
Canada
Goldenrod

Plate 577.
*Echinacea
purpurea*,
Purple
Coneflower

Plate 578. *Leucanthemum
vulgare*, Ox-eye Daisy

Plate 579.
*Anthemis
cotula*,
Stinking
Mayweed

Plate 580.
*Bellis
perennis*,
English
Daisy

Plate 581. *Aster cordifolius* var.
sagittifolius, Arrow-leaved
Aster

Plate 582.
*Aster
divaricatus*,
White Wood
Aster

Plate 583.
*Aster novae-
angliae*, New
England
Aster

Plate 584.
*Aster
puniceus*,
Purple-
stemmed
Aster

Plate 585.
Aster patens,
Late Purple
Aster

Plate 586.
*Aster
linariifolius*,
Stiff Aster

Plate 587. *Aster curtisii*, Curtis' Aster

Plate 588. *Aster acuminatus*, Whorled Wood Aster

Plate 589. *Aster umbellatus*, Flat-topped Aster

Plate 590. *Aster pilosus*, White Heath Aster

Plate 591.
*Aster
lateriflorus*,
Calico Aster

Plate 592.
*Aster
paternus*,
Toothed
White-
topped Aster

Plate 593. *Erigeron pulchellus*,
Robin's Plantain

Plate 594.
*Erigeron
philadelphicus*,
Common
Fleabane

Plate 595.
*Erigeron
strigosus*,
Daisy
Fleabane

Plate 596.
*Galinsoga
quadriradiata*,
Peruvian
Daisy

Plate 597. *Parthenium
integrifolium*, Wild Quinine

Plate 598.
*Achillea
millefolium*
ssp. *lanulosa*,
Yarrow

Plate 599. *Conyza canadensis*,
Horseweed

Plate 600.
Eclipta alba,
Yerba de
Tajo

Glossary of Botanical Terms ❧

Achene. A small, dry, one-seeded fruit.
Actinomorphic. Radially symmetric; capable of being bisected into equal mirror-image halves along two or more planes.
Acuminate. Tapering gradually to a pointed apex.
Acute. Terminating in a sharply-angled point.
Alternate. Situated singly at each node along an axis.
Anther. The pollen-bearing portion of a stamen.
Appressed. Lying flat agianst or nearly parallel to an organ, as leaves to a stem.
Aril. An appendage formed on and more or less enclosing certain seeds.
Ascending. Growing obliquely upward.
Auricle. A small ear-like lobe or appendage.
Auriculate. Having auricles.
Awn. A stiff terminal bristle.
Axil. The upper angle formed between any two organs or structures.
Barb. A short reflexed point.
Beak. A firm, pointed terminal appendage.
Bifid. Two-lobed or two-cleft.
Bilabiate. Two-lipped.
Bilaterally Symmetric. See Zygomorphic.
Bisbilaterally Symmetric. Capable of being bisected into only two dissimilar pairs of equal mirror-image halves.
Blade. The expanded terminal portion of a leaf, petal, or similar structure.
Bract. A modified leaf, differing in size and often in other respects from the foliage leaves, usually subtending a flower or an inflorescence.

Calyx. The outer whorl of the perianth, enclosing other flower parts in bud.

Campanulate. Bell-shaped.

Capillary. Hairlike.

Carpel. A simple pistil or a single unit of a compound pistil.

Category. Any of the universally recognized levels of classification—e.g., family, species, etc.

Cauline. Attached or pertaining to the stem, as opposed to basal.

Chaff. Thin scales or bracts subtending individual flowers in many of the Asteraceae.

Chasmogamous. Not cleistogamous. Denoting a type of flower that opens normally for pollination.

Cilia. Marginal hairs.

Clavate. Club-shaped; gradually thickened toward the apex.

Claw. The narrow, stalklike basal portion of some petals and sepals.

Cleft. Deeply cut, usually more than one-half the distance to the midrib or base.

Cleistogamous. Denoting a type of flower that does not open.

Colony. A group of similar plants growing in close proximity from one root system.

Coma. A tuft of soft hairs on a seed.

Compound. Referring to a leaf, composed of two or more similar parts.

Connate. Describing similar structures that are joined or grown together. (See Connivent.)

Connivent. Describing similar structures that are in close contact but not fused together.

Cordate. Heart-shaped. When used to describe the base only, having a sinus and two rounded lobes.

Corm. A bulblike underground portion of stem for food storage or reproduction.

Corolla. The inner whorl of the perianth, between the calyx and the stamens.

Corona. A crownlike structure appearing in some plants between the corolla and stamens.

Corymb. A broad, flat-topped inflorescence in which the marginal flowers open first.

Crenate. With shallow round or blunt teeth on the margin; scalloped.

Cuneate. Wedge-shaped.

Cyathium. The specialized inflorescence of *Euphorbia.*

Cyme. A broad, flat-topped inflorescence in which the central flower opens first.

Decompound. Referring to a leaf, repeatedly divided into leaflets.

Decumbent. Prostrate at the base but ascending at the end.

Decurrent. Adnate to the petiole or stem and extending downward (referring to leaf tissue).

Deflexed. Bent downward.

Deltoid. Broadly triangular.

Dentate. With sharp, outward-pointing teeth on the margin.

Dichotomous. Forking into two equal branches.

Dioecious. Bearing staminate and pistillate flowers on separate plants.

Discoid. Describing a flower head in the Asteraceae that contains only disk flowers.

Disk. The central portion of a radiate head in the Asteraceae, composed of disk flowers.

Disk Flower. A tubular flower in the Asteraceae, found in discoid heads or in the center of radiate heads.

Divaricate. Widely spreading.

Divided. Cut nearly or completely to the midrib or base.

Drupe. A fleshy fruit enclosing a nut or stone.

E-. Prefix meaning "without"—e.g., eciliate.

Elliptic. Broadest near the middle and gradually tapering to both ends.

Entire. Denoting a continuous unbroken margin without teeth, lobes, etc.

Erose. Minutely and irregularly cut along the margin.

Exserted. Projecting from or extending beyond.

Fascicle. A small cluster or bundle.

Fertile. Capable of performing normal reproductive function.

Filament. The basal, sterile portion of a stamen below the anthers.

Filiform. Threadlike.

Flexuous. Curved alternately in opposite directions; wavy.

Floret. A small individual flower, as in the Asteraceae.

Floriferous. Blooming abundantly.

Foliaceous. Leaflike.

Glabrous. Smooth, without hairs.

Gland. A secretory structure.

Glaucous. Covered with a thin, light-colored waxy or powdery bloom.

Glomerule. A compact cluster of flowers.

Glutinous. Having a sticky surface.

Hastate. Halberd-shaped, with two divergent lobes.

Head. A dense, short inflorescence composed of sessile flowers.

Herb. A plant lacking a persistent woody stem.

Hood. Specifically in *Asclepias*, a segment of the corona.

Horn. In *Asclepias*, one of the beak-like appendages of the corona.

Hypanthium. A cup-shaped enlargement of the receptacle, created by the fusion of sepals, petals, and stamens.

Included. Not exserted.

Inferior Ovary. One that is adnate to the hypanthium, therefore appearing to be located below the perianth.

Inflorescence. The flowering portion of a plant.

Internode. The portion of a stem between two successive nodes.

Involucre. A set of bracts subtending a flower or an inflorescence.

Irregular. Same as Zygomorphic.

Keel. A central longitudinal ridge. Also, the two united lower petals of a papilionaceous corolla.

Lanceolate. Much longer than wide, widest below the middle, and tapering toward the apex.

Lax. Loose or open (referring to an inflorescence).

Leaflet. A single segment of a compound leaf.

Ligulate. Describing a flower head in the Asteraceae that contains only ray flowers.

Limb. The upper, expanded portion of a corolla with fused petals.

Linear. Long and narrow with essentially parallel sides.

Lip. Either segment (upper lip or lower lip) of a bilabiate or zygomorphic structure. Also, the odd petal in the Orchidaceae.

Lobe. A projecting segment or extension.

Lobed. More or less deeply cut, but not all the way to the midrib or base.

-merous. Suffix used to denote the number of parts or divisions—e.g., pentamerous or 5-merous.

Node. A point on a stem where leaves or branches are borne.

Ob-. Prefix signifying inversion—e.g., obovate.

Oblong. Longer than broad with essentially parallel sides (but broader than linear).

Ocrea. A tubular stipular sheath in the Polygonaceae.

Opposite. Occurring in pairs at each node along an axis.

-ose. Suffix meaning "like"—e.g., cymose.

Oval. Broadly elliptic.

Ovary. The basal portion of a pistil.

Ovate. Egg-shaped, wider below the middle (describing a two-dimensional structure).

Ovoid. Describing an egg-shaped solid.

Palate. The raised portion of the lower lip of a bilabiate corolla, within the throat.

Palmate. Radiating from a single point like the outspread fingers of a hand (describing leaflets, lobes, etc.).

Panicle. A compound inflorescence in which the branches are racemose and the flowers pediceled.

Papilionaceous. Describing the structure of a corolla typical of the Fabaceae, composed of a standard, wings, and keel.

Pappus. Collectively, the bristles, hairs, scales, etc., at the summit of an achene in the Asteraceae.

Parasite. A plant which derives most or all of its food from another living organism to which it is attached.

Pedicel. The stalk of a single flower that is part of an inflorescence.

Peduncle. The stalk of a flower cluster, or of a solitary flower not associated with others in an inflorescence.

Peltate. Attached by its lower surface instead of its margin.

Pendent. Hanging downward; drooping.

Perfect. Containing both stamens and pistils.

Perianth. A collective term for the calyx and corolla.

Petal. A single segment of a divided corolla.

Petaloid. Resembling a petal.

Petiole. The stalk of a leaf.

Phyllode or **Phyllodium** (pl. **Phyllodia**). An expanded petiole without a leaf blade.

Pinnate. Arranged featherlike on either side of a common axis (describing leaflets, lobes, veins, etc.).

Pinnatifid. So deeply cut or cleft as to appear pinnate.

Pistil. The central reproductive organ of a flower, composed of ovary, style, and stigma.

Plumose. Appearing plumelike from fine hairs along two sides of a central axis.

Procumbent. Lying flat or trailing but not rooting at the nodes.

Pubescent. Covered with short, soft hairs.

Punctate. Dotted or pitted, usually with glands.

Raceme. An elongate, unbranched inflorescence with pediceled flowers.

Radially Symmetric. See Actinomorphic.

Radiate. Describing a flower head in the Asteraceae that contains both ray and disk flowers.

Ray. A pedicel or peduncle in an umbel.

Ray Flower. A strap-shaped flower in the Asteraceae, found in ligulate heads or on the perimeter of radiate heads.

Receptacle. The expanded apex of a flower stalk, which bears the floral organs.

Regular. Same as Radially Symmetric or Actinomorphic.

Reticulate. Netted.

Retrorse. Bent backward or downward; reflexed.

Revolute. Having the margins rolled toward the underside.

Rhizome. Underground stem, usually horizontal and rooting at the nodes.

Rhombic. Equilateral with oblique angles; diamond-shaped.

Rotate. Widely spreading and circular, with a very short tube.

Sagittate. Arrowhead shaped, with two retrorse basal lobes.

Salverform. With a slender tube abruptly expanded into a rotate limb.

Saprophyte. A plant which derives its food from dead or decaying organic material in the soil.

Scabrous. Rough to the touch.

Scape. A leafless flowering stem arising directly from the ground.

Scapose. Bearing or consisting of a scape.

Scorpoid. Describing a coiled inflorescence.

Secund. Appearing as if borne from only one side of an axis.

Sepal. A single segment of a divided calyx.

Serrate. With sharp, forward-pointing teeth on the margin.

Sessile. Attached directly, without a petiole, pedicel, or other stalk.

Sinus. The space between two lobes, teeth, or other divisions.

Spadix. A spike or head in which the flowers are borne on a fleshy axis.

Spathe. A large bract subtending and usually partially enclosing an inflorescence.

Spatulate. Spoon-shaped; gradually widened to a rounded summit.

Spike. An elongate, unbranched inflorescence with sessile flowers.

Spur. A hollow extension of a petal or sepal.

Squarrose. Having spreading, recurved tips.

Stamen. The pollen-bearing organ of a flower, composed of filament and anthers.

Staminode. A sterile stamen or other nonfunctional structure occupying the position of a stamen.

Standard. The upper petal of a papilionaceous corolla.

Stellate. Starlike, with radiating branches.

Stem. The main upward-growing axis of a plant, on which are borne leaves and flowers.

Stemless. Describing a plant whose peduncles and leaves arise separately and directly from ground level.

Sterile. Incapable of performing normal reproductive function.

Stigma. The terminal portion of a pistil, which receives the pollen.

Stipule. An appendage at the base of a petiole.

Stolon. A horizontal, elongate shoot, either above or below ground, rooting at the nodes or apex.

Striate. With fine longitudinal lines or ridges.

Style. The narrowed portion of a pistil between the ovary and the stigma.

Sub-. Prefix meaning "almost"—e.g., subopposite.

Subtend. To occupy a position below and adjacent to.

Subulate. Awl-shaped.

Superior Ovary. One that is located above and is free of the perianth.

Taxon. Any group of plants occupying a particular hierarchical category, regardless of rank.

Tendril. A slender portion of a leaf or stem, modified for twining.

Tepal. A segment of a perianth in which the sepals and petals are similar.

Terete. Circular in cross-section.

Ternate. In three's.

Throat. In some corollas with fused petals, the point of juncture between the tube and limb.

Tomentose. Woolly, with long, soft, matted hairs; compare Villous.

Trifoliolate. Having three leaflets.

Truncate. With a base or apex appearing as if cut straight across.

Tube. The lower, narrower portion of a corolla or calyx.

Umbel. An inflorescence in which the flower stalks arise from a common point; in a compound umbel this branching is repeated.

Unisexual. Bearing either stamens or pistils but not both.

Villous. With fine long hairs, not matted; compare Tomentose.

Whorl. A circle of three or more structures arising from the same node.

Wing. Any thin, flat, lateral extension of an organ. Specifically, a lateral petal in the Fabaceae and Polygalaceae.

Zygomorphic. Bilaterally symmetric; capable of being bisected into equal mirror-image halves along one plane only.

Bibliography ✤

Adams, Kevin and Marty Casstevens. *Wildflowers of the Southern Appalachians*. Winston-Salem, N.C.: John F. Blair, Publisher, 1996.

Batson, Wade T. *Wildflowers in the Carolinas*. Columbia, S.C.: Univ. of South Carolina Press, 1987.

Campbell, Carlos C., et al. *Great Smoky Mountains Wildflowers*. Northbrook, Ill.: Windy Pines Publishing, 1994.

Dean, Blanche E., Amy Mason, and Joab L. Thomas. *Wildflowers of Alabama and Adjoining States*. University, Ala.: Univ. of Alabama Press, 1983.

Duncan, Wilbur H., and Leonard E. Foote. *Wildflowers of the Southeastern United States*. Athens, Ga.: Univ. of Georgia Press, 1975.

Fernald, M. L. *Gray's Manual of Botany*, 8th ed. New York: American Book Company, 1950.

Gleason, Henry A. *The New Britton and Brown Illustrated Flora of the Northeastern United States and Adjacent Canada*, 3 vols. New York: Hafner Publishing Company, Inc., 1963.

Grimm, William Carey. *The Illustrated Book of Wildflowers and Shrubs*. Harrisburg, Pa.: Stackpole Books, 1993.

Gupton, Oscar W., and Fred C. Swope. *Wildflowers of the Shenandoah Valley and Blue Ridge Mountains*. Charlottesville, Va.: Univ. Press of Virginia, 1979.

Justice, William S., and C. Ritchie Bell. *Wildflowers of North Carolina*. Chapel Hill, N.C.: Univ. of North Carolina Press, 1987.

Klimas, John E., and James A. Cunningham. *Wildflowers of Eastern America*. New York: Alfred A. Knopf, 1974.

Lemmon, Robert S., and Charles C. Johnson. *Wildflowers of North America*. Garden City, N.Y.: Hanover House, 1961.

Luer, Carlyle A. *The Native Orchids of the United States and Canada Excluding Florida.* New York: New York Botanical Garden, 1975.

Martin, Laura C. *Southern Wildflowers.* New York: Random House, 1991.

Mohlenbrock, Robert H. *Where Have All The Wildflowers Gone?* New York: Macmillan Publishing Co., Inc., 1983.

Newcomb, Lawrence. *Newcomb's Wildflower Guide.* Boston: Little, Brown and Company, 1989.

Niering, William A., and Nancy C. Olmstead. *The Audubon Society Field Guide to North American Wildflowers—Eastern Region.* New York: Alfred A. Knopf, 1979.

Peterson, Roger Tory, and Margaret McKenny. *A Field Guide to Wildflowers of Northeastern and North-central North America.* Boston: Houghton Mifflin Company, 1975.

Radford, Albert E., Harry E. Ahles, and C. Ritchie Bell. *Manual of the Vascular Flora of the Carolinas.* Chapel Hill, N.C.: Univ. of North Carolina Press, 1968.

Rickett, Harold William. *Wild Flowers of the United States, Volume 1: The Northeastern States.* New York: McGraw-Hill Book Company, 1966.

———. *Wild Flowers of the United States, Volume 2: The Southeastern States.* New York: McGraw-Hill Book Company, 1967.

Schnell, Donald E. *Carnivorous Plants of the United States and Canada.* Winston-Salem, N.C.: John F. Blair, Publisher, 1976.

Small, John Kunkel. *Manual of the Southeastern Flora,* 2 vols. New York: Hafner Publishing Company, 1972.

Smith, Richard M. *Wild Plants of America: A Select Guide for the Naturalist and Traveler.* New York: John Wiley and Sons, Inc., 1989.

Strausbaugh, P. D., and Earl L. Core. *Flora of West Virginia,* 2d ed. Grantsville, W.Va.: Seneca Books, Inc., 1993.

Stupka, Arthur. *Wildflowers in Color.* New York: Harper and Row, 1965.

Wells, B. W. *The Natural Gardens of North Carolina.* Chapel Hill, N.C.: Univ. of North Carolina Press, 1967.

Williams, John G., and Andrew E. Williams. *Field Guide to Orchids of North America.* New York: Universe Books, 1983.

White, Peter, et al. *Wildflowers of the Smokies.* Gatlinburg, Tenn.: Great Smoky Mountains Natural History Association, 1996.

Wofford, B. Eugene. *Guide to the Vascular Plants of the Blue Ridge.* Athens, Ga.: Univ. of Georgia Press, 1989.

Index ✣

ॐ

Wildflowers of the Southern Mountains *was designed and composed on a Power Macintosh using PageMaker software. Minion Ornaments and Goudy are used for text and display fonts. This book was designed and composed by Kay Jursik and was printed and bound by Milanostampa. The paper used in this book is designed for an effective life of at least three hundred years.*